INTERACTIONS

INTERACTIONS

•••••••••••••

The Biological Context
of Social Systems

Niles Eldredge and Marjorie Grene

Columbia University Press
New York

COLUMBIA UNIVERSITY PRESS
NEW YORK   OXFORD

Copyright © 1992 Columbia University Press
All rights reserved

Library of Congress Cataloging-in-Publication Data

Eldredge, Niles.
  Interactions : the biological context of social systems / Niles
Eldredge and Marjorie Grene.
    p. cm.
  Includes bibliographical references (p. 211) and index.
  ISBN 0-231-07946-X
  1. Sociobiology. 2. Biology—Social aspects. 3. Evolution
(Biology) 4. Social systems. 5. Social interaction. I. Grene,
Marjorie Glicksman, 1910– . II. Title.
QH333.E43   1992
575'.01—dc20   92-9432
                                                      CIP

Casebound editions of
Columbia University Press books
are Smyth-sewn and printed
on permanent and durable
acid-free paper.

Printed in the United States of America

c 10 9 8 7 6 5 4 3 2 1

Contents

INTERACTIONS

Prologue

What can an invertebrate paleontologist and a philosopher possibly have to say about the evolutionary biology of social systems? And hasn't sociobiology already performed the task of welding contemporary evolutionary theory with the biology of social systems—to nearly universal satisfaction, putting aside the residual wrangling over the biological basis of human sociality? These are fair questions, deserving response at the very outset of our joint venture.

Each of us has labored in the vineyards of evolutionary theory. One (the paleontologist) has been concerned with modifying process theory to better fit actual patterns of evolutionary history, especially (though by no means solely) those of the fossil record (e.g., Eldredge and Gould 1972). The other (the philosopher) was initially drawn to epistemological aspects of evolutionary theory (e.g., Grene 1959). To a degree such efforts have identified us, severally, as critics of aspects of contemporary evolutionary theory—and, perhaps more positively, as shapers of alternative viewpoints.

More recently, during the 1980s, we have each become involved in issues of biological *ontology*, namely, the question of what entities actually exist in biological nature, and of what the natures and activities of such entities are. We, along with many colleagues, have concluded that the ontology underlying the "synthetic theory of evolution" (and the particularly hard form of reductive, gene-centered theory that has dominated evolutionary biology since the mid-1960s) is incomplete and partly erroneous (e.g., Eldredge 1985, 1986, 1989; Grene 1987). Our approach has been to propose an expanded ontology, an alternative to, or at least an alteration of, received orthodox views. In particular, we believe that characterization of the *genealogical* and *ecological* or (economic) hierarchies is fraught with implications for the evolutionary process. In no sphere of theory is this truer than in the application of (evolutionary) biology to the understanding of the origin, nature, structure, and function of social systems. In short, sociobiology seems to us in some important respects to be inadequate because it derives from an evolutionary theory that is itself based on a skewed understanding of the nature of biological systems.

These are strong words. The basic thesis of our book is simply stated: *Social systems are historical entities that arise from the reintegration of organismic reproductive and economic adaptations.* That statement may seem disarmingly innocuous, but its context reveals how radical a departure our view takes from the (by now) conventional wisdom of sociobiology. We see biotic entities (populations of several natures, species, monophyletic taxa, and local and regional ecosystems) as arising from the simple activities of organisms. It is the essence of the "new" ontology that reproductive behavior engenders the formation of purely genealogical structures, that sexual reproduction underlies the formation of demes, thence species, thence monophyletic taxa. In contrast, *economic* activity (processes of matter-energy transfer involved solely with differentiation, growth, and maintenance of an organism) leads unequivocally to the formation of other sorts of biotic structures: local (economic) populations ("avatars"), which are parts of local, thence progressively more regional, ecosystems.

The point is that nonsocial organisms are simultaneously part of two separate and distinct hierarchically organized systems. Any organism is a part of a species and of a local ecosystem—and the two have little to do with one another. Local populations of conspecifics (among nonsocial organisms) differ, often sharply, in their economic and reproductive attributes, and the composition of populations (at least in motile organisms) often changes, depending upon

whether purely economic or reproductive activities are being carried out.

In the context of a revised ontology that sees large-scale biotic
systems as either economic or genealogic in nature, the conclusion
that social systems are entities arising from a reintegration of organismic economic and reproductive activity is rather obvious—and
should not be, at first glance, particularly controversial. We have
summarized our ontological scheme of biological systems pictorially
in figure 1. The diagram both illustrates our thesis (in a very general
way) and illustrates our departure from contemporary neo-Darwinian (or, as we shall call it, "ultra-Darwinian") perspectives. For though
our thesis does appear, initially, as a description and analysis that
need not be taken as a direct counterproposal to the theses of
sociobiology, there are ingredients that ultra-Darwinians—and hence
sociobiological theorists— will not like.

Let us, for a moment, assume the accuracy of the ontology depicted in figure 1. As we shall explain more fully in the text, evolution in general is the historical outcome of interaction between
elements of the ecological and genealogical hierarchies. In particular, natural selection (following Darwin 1859 rather closely) is the
relative effect wrought by differential economic success on reproductive success within a local population of conspecific organisms.
Natural selection is the between-generational vector connecting the
economic side with the genealogical side of the ledger, biasing the
transmission, and hence the long-term accumulation, of genetic
information. Genealogical systems (local demes and, more stably,
species and monophyletic taxa) are historical entities of packaged
genetic information. They are (among) the results of the evolutionary process. In contrast, economic biotic entities (such as ecosystems) result from moment-by-moment economic interactions among
component parts.

It is canonical in contemporary evolutionary theory, beginning
perhaps with Williams's (1966) *Adaptation and Natural Selection* and
culminating with the explicitly extreme version of Dawkins's (1976)
The Selfish Gene, that evolution boils down to a competitive game to
maximize representation of one variant version of genetic information over another in the succeeding generation. Fitting hand in
glove with optimization strategies in behavioral ecology, the game is
for an organism to maximize its genetic representation in the descendant population. Whether seen from an organismic perspec-

FIGURE 1.

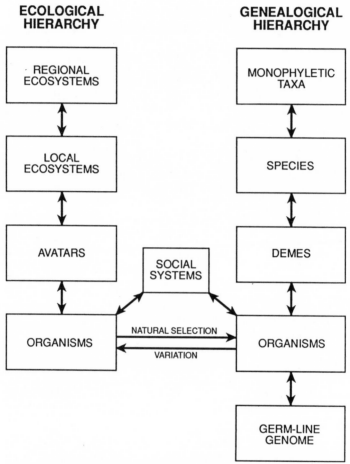

A visual schematic of the Ecological and Genealogical Hierarchies. Boxes represent various categories of entity. Entities at any level are composed of entities at the subjacent level. Arrows depict aspects of upward and downward causation (see, e.g., Eldredge and Salthe 1984; and Vrba and Eldredge 1984 for discussion). Every single organism is simultaneously a member of an ecological and a genealogical hierarchy. Social systems are unique, above the level of single organisms, in that they represent fusions of economic and reproductive organismic behaviors and functions.

tive—or from the more reductive gene's-eye view—organismic behavior generally has come to be regarded, at base, as being concerned exclusively with the maximization of reproductive success.

This postulate represents a subtle, yet profound, change in the
perception of what natural selection actually is. We have said that
we see natural selection more or less as Darwin did, as a simple
accumulator, an imperfect yet effective recorder of what seems to
work best in a finite world of varying organisms where offspring
tend to inherit those features that confer relative economic success
on the parents. The description of nature proferred by ultra-Darwinians—and maintained without change in the sociobiological
canon—is very different. In that view, organisms actively seek to
maximize their reproductive success—a proposal with which we are
comfortable, since it is easy to understand how selection would have
acted to produce this result. But ultra-Darwinians go further: *all*
aspects of living systems are to be understood, ultimately, as an
outgrowth of competition for reproductive success. Economic competition, then, becomes just an expression of the underlying cause:
reproductive competition. Darwin saw that economic competition
leads to differential reproductive success. Modern evolutionary biology sees economic competition as a direct reflection of the *real*
competition: for reproductive success. Under this rubric, natural
selection is transformed from a passive accumulator to a dynamic in
nature, responsible, at bottom, for the organization of all manner
of biotic entities: species, local ecosystems, and, most of all, social
systems.

Our view of the ontology of biotic systems sees the effects of
economic and reproductive activity as primarily separate. When we
turn to social systems, we see economic and reproductive activity
integrated (in interestingly disparate ways when one compares, e.g.,
colonial invertebrates and hymenopteran insects, on the one hand,
and [most] social vertebrates on the other). In contrast, the sociobiological literature is replete with the assumption that the actual
structure of any given society is a reflection of the competition for
reproductive success—definitely a reformulation into active mode
(if not a downright perversion) of the genuine evolutionary principle of natural selection. In taking a historical principle for an active
dynamic, sociobiology in effect says that reproductive biology is the
basis of social systems. In contrast, our perspective grants equal
footing to economics and reproduction, a point of departure that is

6 PROLOGUE

extremely useful when we finally confront the biological aspects of human sociality.

Returning, then, to figure 1, one can describe the system depicted from any point of departure. That modern evolutionary theory in large measure prefers to enter at the reproductive organismic level (if not the germ-line gene perspective of an unregenerate Richard Dawkins) is in itself no crime. Insisting, however, that everything can be reduced to that ontological level simply because the perspective originates there is, to our mind, to distort the description of the larger system, with all its disparate elements. We strive here simply to argue that the system in its entirety exists, that it can be entered on either side at any level, and that, in any case, where one's perspective begins should not diminish the relative importance of every other element in the rich tapestry of hierarchically organized biotic systems.

Our book, then, is an extended argument on biological ontology. Only the final two of our seven chapters discuss social systems *per se*. Our first chapter locates our project both *vis à vis* sociobiology and within the current heightened pluralism of contemporary evolutionary theory. In chapter 2 we examine the relation of Darwinian theory to pre-Darwinian scientific explanation of biological phenomena. We then turn to a consideration of Darwinian explanation generally—which allows us to take a closer look at the nature of ultra-Darwinian explanation.

In chapter 3, our concern is to establish the two great classes of biological activity: economic (matter-energy transfer processes) and reproductive. In a phylogenetic context, as organisms become structurally more complex, the gulf widens between reproductive and economic structure (and function), and the subdivision of labor among various discrete economic tasks increases. Chapters 4 and 5 examine the nature of large-scale biotic entities of which organisms are parts. Chapter 4 (on genealogical systems formed through organismic reproductive activity) contains a lengthy treatment of species as the paradigmatic example of a pure genealogical system that offers similarities to and stark differences from social systems. In chapter 5 we describe in detail the elements of the ecological hierarchy, which, with each element arising and being maintained solely through the economic interactions among its components, likewise reveals aspects pertinent to understanding the biological nature of social systems.

Chapter 6 then fully develops the argument that social systems are hybrid systems, roughly at the level of the local population (avatars and demes, of the ecological and genealogical hierarchies, respectively). They are unique as biotic systems (i.e., systems composed of organisms) in that economic and reproductive organismic functions are deeply intertwined—albeit in strikingly different ways in different monophyletic taxa.

And, finally—almost inevitably—we apply our perspective to human sociality. We note that sex is significantly decoupled from reproduction in human behavior and argue that it forms a complex bridge between human economic and reproductive behavior. We outline no new theory of the biology of human sociality. But we do see a biological basis for understanding human sociality as arising at least as much through economic as through reproductive behavior—a point, of course, long grasped in anthropology and sociology but lost in the rush to apply wholesale the principles of ultra-Darwinian evolutionary theory to human sociality *via* sociobiology.

That social systems have evolved cannot be doubted. That economic and reproductive behaviors are the dynamics that hold non-human social systems together is also indisputable. That such biological processes have much to do with human societies must also be true—although the human capacity for symbolic activity clearly locates any attempt to understand human sociality on a plane not occupied, as far as anyone knows, by any other species (though partially anticipated in some other primates). Human sociality must of course be consistent with what is known of the biological foundation of social systems generally. This does not mean, however, that all elements of human sociality must arise as simple functions of basic biological activity. At the very least, we hope to have provided a more balanced picture and a more accurate map of the nature of biotic systems generally, the better to grasp how organismic behavior leads to the formation and maintenance of all manner of social systems.

CHAPTER ONE

Contemporary Evolutionary Perspectives

After Reductivism

Twenty or more years ago the so-called problem of reductivism loomed large in the thought of philosophers of science and of biologists—at least there was an extensive and lively literature on the subject. Better, perhaps, reduction was a problem for those who feared it and a claim for those who welcomed its (alleged) success. Biology, some boasted, could be, or had been, reduced to physics and chemistry. No way, others countered: the laws of physics and chemistry oversimplify, and the richness of biological phenomena necessarily eludes their grasp.

There was a parallel situation in evolutionary theory, although here for a while a particular form of reductivist thought, genic selectionism, made itself heard in conspicuously strident tones. Evolutionary theory, some very authoritative writers claimed, was reducible to—indeed, had been reduced to—differential gene frequencies. Poor old Darwin and other quaint Victorians, we were told, thought of a literal struggle for existence, "nature, red in tooth and claw" (a phrase in fact coined by Tennyson some years before the publication of the *Origin of Species*). We now know, it was pronounced, that it's all a question of ratios of genes, quantitatively expressed. No need to worry any longer about anything so sentimental as struggle, so gross as behavior, so everyday as lions and tigers, walruses and dinosaurs, and what they *do* to one another. It's only the algebra of relative gene frequencies that can be statistically recorded, so it was said, therefore, we were told, that's what evolution *is*. This is a kind of mathematicizing analogue of Hacking's tag—"If you can spray it, it's real" and by seeming implication, otherwise it isn't—if and *only* if you can spray it, it's real. In this case, if and only if you can quantify it, it's real. That has been one of the hopes of our scientific, or scientizing, tradition ever since it started. The operationalist imperative so neatly expressed by Hacking is, of course, another motif of the same tradition, but not the one we are concerned with here. The reductivism of biologists and philosophers of science in recent decades has been not so much operationalist or phenomenalist (like old-fashioned positivism) as it has been mathematicizing, and, as we shall see later, atomizing (or molecularizing) in its tendency.

Of course, there were always a few who resisted obeisance to the all-conquering gene. What concerns us here, however, is that recently, and fortunately, the oversimplifying attitude of the reductivist era has become less prominent. Biologists and philosophers of science are paying more attention to the complexities of the entities they are trying to understand. Even where reduction takes place, it is within limited domains and in the context of particular, less global problems. In general, in thinking about the "foundations" of biology, both biologists and philosophers have become less given to single-minded abstractions, less global, and thus less comprehensively reductionist in their approach (e.g., Kitcher 1984; Burian and Grene 1990). Recent advances in molecular biology make this change of mood especially clear. No one anticipates, as was the case in the heyday of the "biochemical revolution," that biology will be replaced by the new and more analytical disciplines, or perhaps only the new and more analytical techniques, now available at the molecular level. For evolutionary theory in particular, molecular biology seems to provide still more experimental underpinning and theoretical sophistication than had been offered by Mendelian, and in particular by population, genetics: collaboration, not conquest, seems to be the order of the day (Burian and Grene 1992).

This welcome shift of emphasis can be seen throughout evolutionary biology. There have been and continue to be a number of conspicuous efforts to counteract the genic reductivism of earlier decades, to enrich and to complexify (without, we hope, a loss of clarity) the basic concepts and interests of evolutionary theory. It is within that context that the present undertaking should be placed. We will be concerned here with one consequence, as we see it, of one such more complex and less abstract version of the traditional (basically Darwinian) theory of evolution: the consequence for the evolutionary account of the origin and nature of social systems of a hierarchically expanded and analytically enriched evolutionary perspective.

Since this expanded and enriched evolutionary theory is in process of development—and since both of us have so far been more concerned with that development than with the problem of social systems as such—much of the text that follows will trace such details of the new two-hierarchy model as are necessary in order to locate social systems (in particular, the social systems of animals) within evolutionary history.

As our title indicates, moreover, our perspective is wider than social systems as such. We are adopting, and adapting for our purposes, the distinction introduced by David Hull (1980) in his now-classic paper "Individuality and Selection"—to which we will refer again shortly, and recurrently, in the course of our argument—between *replicators* and *interactors*. According to Hull's account, replicators and interactors interact to produce lineages and thus evolution. It is plain that interactors interact not only in societies—with their conspecifics—but with other entities as well. They are influenced by (and influence), for example, predators and prey, neither of which are "normally" conspecific, and such abiotic environmental variables as climate, water supply, quality of air, and so on. Clearly such interactions transcend the limits of what is usually understood as a social system.

Further, it is not interactors as such that give rise to evolutionary lineages; but neither is it simply replicators, in abstraction from environmental and therefore interactive contexts. On the contrary—we must look to the *interactions* of interactors *and* replicators to find the patterns of evolution, including, of course, the evolution of the social systems we are seeking to understand.

Interactors, interactions: this may seem a cumbersome near-duplication of terminology, but it is a crucial one. Robert Brandon's (1990) model of adaptive evolution, which differs somewhat from the standard Darwinian story, brings out the point here with unusual clarity. Brandon, relying as we are on the distinction made by Hull, sees selection as occurring among interactors, that is, among phenotypes. But this is not yet evolutionary selection. When such (phenotypic) selection issues in differential reproduction—that is, when it has an effect on replicators—then and only then do we get evolution, with lineages exhibiting descent with modification. Thus Brandon echoes evolutionists like Mayr (1978) and Wright (1967) in arguing that we have a two-step process. However, instead of considering random variation as the first step and the selection of heritable differentially "fit" variants as the second (as Mayr, Wright, and others have) Brandon sees phenotypic selection as the first step and differential reproduction of such selected variants as the second. In this account it is plain, perhaps plainer than in the usual story, that while interactors interact with (and in) their conspecific, biotic, and even abiotic environments, it is the interaction of those interactive agents or events with replicators that results in evolution.

We will focus on interaction in this broad and divided context. We will look first at biotic systems as systems of reproducers on the one hand and interactors on the other, that is, at the genealogical as distinct from the economic adaptations of living entities, and vice versa; we will then look at the interactions between those two systems that issue in descent with modification, that is, in evolution. It is only when we have, somewhat laboriously, constructed this complex context that we can try to envisage the origin, and the *raison d'être*, of animal sociality. Thus our readers must be prepared to forgive us if, like Hume, we "beat about the neighbouring fields" on our way to a not-quite-asymptotically-receding goal.

Before we embark on this undertaking, however, we should situate our project somewhat more precisely in the current literature on the evolution of social systems and more generally in the literature of evolutionary theory. Among many current fashionable possibilities, we should make it clear what we are *not* about to try to do! So we shall try to sort ourselves from others who have attempted, or are attempting, to produce evolutionary accounts of social systems and also from others who are proposing expansions of or supplements to Darwinian evolutionary theory.

Social Systems in Evolutionary Perspective: The Claims of Sociobiology

That evolution happened—that the vast range of living phenomena observable on this planet today have come about "historically," from remote beginnings conspicuously different from contemporary events and entities—this is the tacit presupposition of all research and discourse in every corner of the biological sciences. Even systematists impatient with the Darwinian orthodoxy and eager to ignore evolutionary theorizing in their work, even such transformed systematists have to admit—especially given the enthusiasm of some of them for panbiogeography—that the flora and fauna of the earth have changed as the earth and seas have altered. If this common premise is granted, however, evolution must be seen to embrace not only the structures but also the behaviors of organisms. Social systems—societies and the entities of which they are composed—have histories, too. If all organisms are descended from a single ancestor, it follows that diversity of behaviors shares with anatomical and physiological diversity an evolutionary origin and, in general, it would appear that more complex social systems develop from relatively simple systems.

Let us take just one case here to illustrate a very complex and variegated range of subject matter: a sweat bee studied by George Eickwort of the Cornell entomology department (Eickwort 1986: cf. West-Eberhard 1979). It is the amazing eusociality of wasps or bees on which, in the first instance, the doctrine of sociobiology was founded: cost-benefit analysis, borrowed from economics, was applied to the social life of insects, and this model was then extrapolated to explain, in terms of differential gene frequencies, the evolution and significance of all animal behavior. But that form of social life is itself, it appears, a development from less fully eusocial life histories. From a painstaking study of the sweat bee *Dialictus lineatulus,* Eickwort concludes that this organism represents a primitive stage of eusociality:

> Nests begin as solitary or communal associations. Some of them, briefly, develop into eusocial colonies, but for most of the summer the colonies are semisocial. In one place and in one season, all the major categories of social behavior are represented. Selection is not acting primarily on the eusocial aspects of *D. lineatulus,* it is the "less advanced" communal and semisocial aspects that have enabled the

sweat bee to become a successful species in the lawns of eastern North America. (Eickwort 1986:753)

Eickwort's study illustrates well the subtlety and diversity of data and of concepts that can be applied to the evolutionary study of social behavior, as distinct from the more traditional study of the evolution of structural differences in the biota.

Very well: social systems certainly can be, and have already been, widely and conspicuously, the subject of evolutionary study. Before we start on what we consider a new attempt, on a broad (shall we say "macroevolutionary"?) scale, to place social systems in their evolutionary context, we must first address the claim that our work has already been done for us. For it does clearly appear to many that the assimilation of social phenomena to evolutionary biology has already been accomplished. A decade and a half after the publication of E. O. Wilson's *Sociobiology* (1975), many believe that this aspect of evolution has been dealt with, finally, once and for all. Indeed, the explication of the origin and nature of animal and human social systems in evolutionary terms has been, its founders and disciples believe, the great achievement of sociobiology: it remains, they hold, only to elaborate the consequences of this fundamental breakthrough. For sociobiology has not only addressed social systems in explicitly evolutionary terms, it has done so in terms effectively identical to, or at most providing only a slight extension of, conventional neo-Darwinian evolutionary theory, with its firm genetic basis. Thus sociobiology is widely regarded as providing at last a solid approach to the evolutionary basis of social behavior.

In many special areas and for many specialized problems, this claim seems justified—especially if we interpret the name *sociobiology* in a rather wide sense. For notwithstanding the notoriety that virtually all attempts to "explain" human behavior in sociobiological terms invariably attract, even its severest critics may concede that sociobiology, or its less genetically based cousin, behavioral ecology, has gone far in clearing up diverse mysteries of animal behavior— "altruism" perhaps foremost among them. It is not our purpose to take issue with such contentions, although it should be remarked in passing that the use of "altruism" in this literature raises grave conceptual problems. Sociobiologists take a concept from a human context, apply it to, say, insect behavior, and then reapply the denatured concept, with a new meaning, to the behavior of very different animals—mammals, primates, human beings. Nevertheless, the cant about "altruism" does reflect a problem Darwin had worried about: natural selection works for the "good" of each being; how can selection favor behavior for the "good" of another, not oneself? That is a problem inherent in Darwinian thinking, and sociobiology has performed a neat end run around it. Indeed, it seems inarguable that, particularly via the concept of inclusive fitness (Hamilton 1964a, 1964b), students of animal behavior have achieved a firmer evolutionary grasp of all manner of behavioral items, from altruistic behavior through the zoology of senescence. Much excellent work has been and continues to be done in behavioral ecology in the study of the behavior of particular kinds of organisms; see, for example, the representative collection by Krebs and Davies (Krebs and Davies (2d. ed.) 1984; cf. Stephens and Krebs, 1986). True, since it lacks the genetic foundationalism of sociobiology, *sensu stricto,* behavioral ecology in general and its chief modeling device, optimization theory, in particular, are distinguishable from sociobiology as such. But their experimental methods and empirical results are closely parallel, if not identical. So it seems fair to credit something loosely called "sociobiology" with much of the impressive and proliferating work in this area. We shall look briefly later at the Darwinian foundation of such work.

At the same time, however, the best experimenters and thinkers in this field admit its own limitations—at least in their more thoughtful moments. Thus Oster and Wilson (1978), while using optimization theory—the characteristic instrument of sociobiological research—as their tool for studying caste systems in social insects, present a careful "Critique of Optimization Theory," in which they conclude: "In our opinion, the way forward in evolutionary theory is not through the formulation of global statements about the evolutionary process, but through the prudent choice of paradigmatic examples that permit the role of natural selection to be analyzed with unusual clarity" (Oster and Wilson 1978).

Now, admittedly, our current enterprise, including as it does some pretty global speculations about the patterns of evolution, presumably counts as one of those decidedly not marking the "way forward." But how paradigmatic are Oster and Wilson's examples, primarily the evolution of caste in the social insects? How far can such cases be extrapolated, as sociobiology seems to claim, to the whole process of evolution?

Other writers, like Krebs and McCleery (1984) also elucidate the presuppositions that limit the scope of work in this field (for replies to criticisms, see Stephens and Krebs 1986). John Emlen (1988) goes so far as to see optimization theory as purely "operational," having no claim to realism. Indeed, the self-critique of Oster and Wilson echoes this attitude, even if these authors seem to take back with one hand what they have granted with the other: the mathematicization inherent in their methods may lead them far from biological reality; however, since their work is more precise than other approaches, it is, so to speak, the only way to go. One is reminded of Levins's much-cited paper on the use of models in biology (1966). The precision, scope, and realism of biological models, he argues, must be balanced against one another. And where realism fails, there seems little chance that the model in question can be legitimately extrapolated to the widest range of evolutionary phenomena either in space or time.

Perhaps the limits of this kind of work—not only of the genetically based formulas of self-titled sociobiology, but also of the more phenotypically oriented field of behavioral ecology—are best stated by John Endler. As he puts it, optimization theory is an "equilibrium theory" (1986). Taking as given the whole sweep of evolution, itself understood in strict Darwinian (or neo-Darwinian) terms, it explains in detail the development of certain behaviors—for example, foraging or mating behavior or parental care in species A, B, or . . . N, once that overall evolutionary process has occurred (see, e.g., Winkler and Wilkinson 1988). There *are* barn swallows, hummingbirds, and sticklebacks; given their existence and their origin on the ground of a general theory of evolution by natural selection, investigators ask: how did they come to behave vis-à-vis prey or mates or offspring in this particular way?[1] Of course, even in the study of behavior, the reductivist attitude of programmatic sociobiologists may sometimes lead them to ignore or to undervalue alternative, less atomizing approaches to the study of social behavior (Gordon 1988, 1990). Nevertheless, for particular cases something akin to a sociobiological approach to the study of an animal behavior has surely produced and may yet produce beautiful work, and we do not of course on principle have any quarrel with it (although, as we have just suggested, the question of extrapolation to all evolutionary phenomena on every scale remains debatable).

However, in the broader sweep of the sociobiological explanation of the origins of social systems, we find—as have many of its critics— a double one-sidedness (if that is possible), a narrowness that is sometimes reflected also in the sociobiological treatment of particular kinds of social behavior. The sociobiological application of neo-Darwinian principles to the explanation of social behavior lies in the straightforward reduction of complex phenomena such as social systems to the standard form of among-organisms/within-populations differential reproductive success. Such discourse entails two often inextricably interwoven weaknesses.

First, it is invariably and almost inevitably teleology-ridden: we measure the "goodness" of an adaptation by assessing its effects on reproductive success, concluding what we have all along assumed—that adaptations of all sorts are "for" the perpetuation and transmission of their underlying genetic instructions (the point, e.g., of Dawkins' [1976, 1982] "selfish genes"). This focus on who or what benefits has not been lost on critics of sociobiology, who have pointed our the "hyperadaptationism" that links sociobiology directly with the mainstream of modern evolutionary biology. (Those who embrace the theory, of course, take this for one of its virtues [Ruse 1986].) Gould and Lewontin's (1979) criticism of Barash et al.'s treatment of "cuckoldry" in bluebirds is now classic; Kitcher (1985b) has reviewed these and other definitive criticisms, and we need not repeat them here. The point is that sociobiological explanation, in focusing, as single-minded Darwinian explanation often does, on differential reproduction as *the* end from which all evolutionary explanation derives its meaning and content, becomes what we may call pseudoteleological in its style and import. All Darwinian theories, of whatever style, are in the last analysis adaptive theories—and adaptation is a what-for phenomenon. But the important point (to which we shall return in later chapters) is that while Darwinian theory explains adaptive phenomena—structures, functions, and behaviors that are good for something in some particular way in some particular context—its explanatory *force* is not teleological, but deterministic-times-stochastic. The theory of natural selection, which is the core of Darwinism, explains how organisms are altered over time as the consequence of certain interactions between organisms and their environments, with subsequent effects on reproduction such that future generations differ from past ones in significant ways. But this is a classic when-then explanation, not a teleological one. It says not that these means happened for the sake of this end, but rather that because certain previous circumstances permitted certain other organism-environment interactions, this adapted end state came about.

Adaptedness, therefore, enters into Darwinian evolution at the level of the *explanandum,* not the *explanans.* An elegant account of this situation was given in Robert Brandon's (1981) little essay, "A Structural Description of Evolutionary Theory," which placed adaptive phenomena very neatly within the conceptual structure of Darwinian theory (see also Brandon 1990). What we want to stress here is a corollary of that analysis: namely, that while the phenomenon of adaptedness, which natural selection is invoked to explain, is certainly telic—characterized by means-end relations—the explanation itself is clearly causal, *not* teleological. Standard sociobiological explanations tend to conceal this fact. Thus, insofar as they produce their characteristic just-so stories, such accounts may rightly be condemned for their pseudoteleology. They put the teleology in the wrong place, and once they have it there, they use it irresponsibly.

At the same time, in such irresponsible sociobiological explanations pseudoteleology is combined with a tendency toward the biological equivalent of a global and naive atomism. It is a small step in modern evolutionary thought from "sharp teeth are for efficient capture of prey" to "sharp teeth are for perpetuation of their underlying genetic information." Sociobiology comes close to having us conclude that social behavior is not only for maximization of reproductive success, but for maximization of reproductive success at the genic level only—the modern biological equivalent of the traditional belief that realities are always to be explained with reference to the existence, and in this case the perpetuation, of their least parts. Thus in Darwinian terms—or better, in what we shall call ultra-Darwinian terms—organisms, and, *a fortiori,* the social systems composed of aggregates of particular organisms, are simply vehicles for the carriage of genes from one generation to the next.

Doubtless for carriage makers their products were aesthetically and technically marvels of design. But at bottom they were simply devices meant to transport, which did successfully transport, members of the aristocracy from boudoir to ballroom and back again, untouched by the muddy streets and the malodorous rabble who walked through them. Similarly, the dazzling array of multicellular life that has proliferated on this planet, from the earliest algal forms to mice and men, is but a collection of vehicles for the transmission of genes from then to now and from now to a new then. Just as for the ancient atomists there were only atoms and the void, so for Darwinian-Dawkinsian evolutionists there are, in the last analysis, only genes and the temporal sequence along which replication impels them. Yet organic systems—from molecules up through organisms and beyond, to ecosystems, species, and societies—are not often concerned solely, or even primarily, with the transmission of genetic information. For there is an *economic* theme to nature that is at least as basic to and characteristic of living systems as is the maintenance and transmission of genetic information. That economic theme is concerned, at its atomic level, with matter-energy transfer, or, at the organism level, with the development, growth, maintenance, and ultimate death of the soma. Reconsideration of the sorts of biological entities that actually exist in nature and take part in the evolutionary process (Eldredge 1985, 1986) has led to both a revision and an expansion of the ontology of the classical "synthetic" theory of evolution. And recognition of the hierarchical organization of various sorts of biological entities has helped clarify the nature of those entities—thus providing a framework for a more complete, hierarchically based evolutionary theory and, in addition, a more balanced context for understanding what social systems are and how they evolve.

This theme will also be elaborated in later chapters. Here we simply emphasize the self-limiting character of sociobiological explanation when it is honest and its pseudoteleological, cryptoatomistic character when it spreads itself to embrace all evolutionary phenomena.[2]

Finally, from a philosophical point of view, there is an even more fundamental objection to the strict (reductivist and extrapolationist) sociobiological program. What philosophers call hard determinism, and even more conspicuously an atomizing hard determinism, refutes its own claim to constitute knowledge. Sociobiology is allegedly science; science rests on evidence. But the statements made by strict sociobiologists are determined by their genes, and our resistance to their statements is determined, just as firmly, by our genetic packets. So there is no evidence, no reason to believe one way or another— only blind cause and effect sequences. Another way to state this thesis would be to show, as did John Searle (1978) in a brilliant essay, and Marshall Sahlins (1976) in an equally powerful (if partly flawed) argument—that sociobiology can give no account of *intensionality:* that is, that human discourse is *about* something (cf. Grene 1978). There is a more than ample philosophical literature on this topic; this is not the place to add to it. For sociobiology, the upshot is the same as from the determinism argument: if there is no intensionality, sociobiology is not science but a tale . . . never mind the intervening four words! . . . signifying nothing. Again, we believe that a hierarchically developed evolutionary theory can avoid this self-contradiction.

That is all by the way. Our primary concern here is certainly not to assess once more the achievements or ambitions of sociobiology as such (that task has been, we believe, definitively accomplished by many others (see, e.g., literature in Kitcher 1985b; Gregory, Silvers, and Sutch, eds. 1978), nor to evaluate the varied experimental and theoretical work accomplished in various areas under the influence of some of its concepts or some of its methods. Rather, we are concerned to approach the problem of the origin and nature of social systems from the perspective of a richer and, as we see it, more adequate evolutionary theory. This is not—as we will acknowledge in more detail later—an anti-Darwinian or even a non-Darwinian theory. We recognize, as all Darwinians do, the central role of adaptive phenomena in the story of life on this earth. But we deplore—as will also become clear—the hyperteleologism combined with hyperreductivism of much evolutionary thinking, past and present. So from now on any reference to sociobiology will be purely incidental to our own argument; on the matter of principle, let others speak for us.

The Current Scene in Evolutionary Theory: Expansions and Modifications of Darwinism

If we intend to speak of the origin and character of social systems from the perspective of an expanded evolutionary theory, we must look first at some of the modifications and elaborations of orthodox evolutionary theory currently debated in order to place our project in this new landscape.

The decades from 1930 to 1960 were dominated by the development and celebration of the evolutionary synthesis. Some of the classics of the synthesis were modest and somewhat tentative in their initial statements, but later revisions expressed a more dogmatic, sometimes even propagandistic, confidence in the definitive, once-and-for-all character of the reconciliation of Darwinian and Mendelian theories that justified the designation "synthetic theory" or "evolutionary synthesis." This change has been discussed by a number of writers, notably Gould (e.g., 1983) and Eldredge (1985; cf. Grene 1981). Many, though not all, of the contributions to the centennial volumes edited by Sol Tax (1960) also expressed this celebratory tone. It seemed to many evolutionists, and also to biologists who relied on, without working in, evolutionary theory, that a definitive formulation of (effectively) Darwinian theory had been achieved. For a brief period there was an aura emanating from evolutionary literature rather like that surrounding eighteenth-century discussions of Newtonian physics.

That day is over. Challenges to the short-lived orthodoxy abound. While we will have occasion to discuss some of these developments and to deliniate our own proposal for reform, it may be appropriate here to describe briefly various themes in the current literature and to illustrate what we consider some of the more promising directions of innovation in evolutionary theory.

The major theme of evolutionary theory in the Darwinian tradition is, of course, adaptation: Darwinian theory has been concerned—as was neo-Lamarckism in its time—with slightly different organisms tracking differences in their environments. Both before and after the great days of the synthesis, however, less adaptively focused theories of evolution have been put forward. In the first decades of this century, indeed, such theories proliferated to the point that the major historian of the synthesis, W. B. Provine (1989), proposes to rechristen it "the evolutionary constriction." Given the achievements of the synthesis that Provine himself (1971, 1985; Mayr and Provine 1980) has done so much to record, that seems a little unfair. But it certainly is the case that, say, up to the 1930s, rather sweeping views of evolution were proposed that found the propelling force of the evolutionary process not in the relation of organisms to their environments but in some inner compulsion of "life" itself. There was Berg's "nomogenesis," for example, Osborn's "aristogenesis," and a cluster of "emergence" theories, which made some kind of drive to "higher" forms, rather than adaptation, the central theme (see Bowler 1983).

It was in the 1930's, too, a few years after the definitive publications of Fisher, Haldane, and Wright, almost contemporary with the first edition of Dobzhansky's *Genetics and the Origin of Species* (1937) and a few years before Simpson's *Tempo and Mode* (1944) and Mayr's *Systematics and the Origin of Species* (1942), that Robson and Richards published their *Variations of Animals in Nature* (1936), which was vigorously skeptical of natural selection as *the* force in evolution. As late as 1953 the eminent British geologist H. L. Hawkins (1953) delivered an address in which he reiterated the theme of "racial senescence." Organisms like *Gryphaea* or the Irish elk were supposed to illustrate the way species, like individual organisms, develop, atrophy, and die. And there are other cases of "heretics" in this period, such as Goldschmidt (1940) and Willis, (1940) who, though not widely accepted, at least had to be seriously reviewed and refuted. In the English-speaking world at least, however, all these alternatives were soon silenced. In all but a few inconspicuous corners, a special style of Darwinian orthodoxy appeared to hold sway (see Gould 1983 and Grene 1981). Even in this period of the hardened synthesis, there were, to be sure, differences and debates: the split between the classical and balance view in population genetics or the Fisher-Wright disagreement, which generated two rather different traditions in the training of evolutionists (Provine 1985). But from the outside, there was for a time an almost monolithic appearance of unanimity.

At present, in contrast, a surveyor of the literature would find a much more diversified scene. It has been remarked that "the synthesis is effectively dead" (Gould 1980a) and that a "dyssynthesis" is needed (Antonovics 1987). One of the founding fathers of the synthesis itself, G. L. Stebbins (1987), has announced that it was in fact only "the synthesis of mid-century," which is now giving way to a new and wider synthesis, drawing important lessons from advances in molecular biology and focusing on phenomena of "comparative evolution." There are also more vehement desynthesizers, who reject on principle any Darwinian ("adaptive") approach to evolution. Let us locate ourselves briefly in this new arena.

If we dissociate our project from the most notorious Darwinian, or ultra-Darwinian, attempt to explain social life in evolutionary terms, we wish just as emphatically to put aside any affiliation with the numerous anti-Darwinian evolutionary scenarios that are again proliferating. That molecular biology in general and the neutral theory in particular have a contribution to make to the general physiognomy of evolutionary theory we do not for a moment deny; but the neutrality of mutations does not undercut the general patterns of adaptive evolution, and it has, in any case, no visible relevance for our topic. Other anti-Darwinian theories, such as those based on irreversible thermodynamics or "N.E.T.," lie even further from our interest (Weber, Depew, and Smith 1988). So far, it seems to us, these writers have not made a convincing case for deriving laws of evolution, ahistorically, from the laws of physics. If we talk a lot about matter-energy exchange, we are talking of the economic life of organisms, or of ecosystems, which are composed of organisms. We are not suggesting a radical reduction of biology to theoretical physics, nor, be it said once more, a radical rejection of Darwinian principles in the description and explanation of life's history. We seek to enrich and expand the tradition, not to reject it.

There are, however, other, less radical reforms of classical Darwinism with which we find ourselves to some extent in sympathy. Primarily, as will be evident throughout, we subscribe, if in our idiosyncratic way, to the aim of a hierarchical expansion of evolutionary theory that recognizes systematically nested entities at a number of levels and (as we see it) in two distinctive dimensions, the genealogical and the ecological, or the informational and the economic. Indeed, our hypothesis about the origin of social systems rests on that distinction between two evolutionary hierarchies.

The classic statement of that division may be found in Hull's "Individuality and Selection" (1980), to which we have already referred, in which he introduced the distinction between replicators and interactors. Hull himself associates that essay with the pronouncements made a few years earlier to the effect that species are individuals (Ghiselin 1974a; Hull 1976). The species = individuals literature, however, is entangled with numerous logical and philosophical confusions that we hope we can avoid. Two of them we mention briefly here.

First, logic is a well-established formal discipline, either a branch of mathematics or its foundation, in which the alleged pervasive distinction between (eternal, single-propertied) classes and individuals is all but unintelligible. Hull has always (though sometimes more clearly than others) made it plain that he is simply making a distinction he wants to make, as, e.g., between whale and Gargantua. But his colleagues in biology, both before and after his first papers on this question (Hull 1976, 1978) have taken this to be a distinction of or in logic. Species, it is said, are now acknowledged to be "logical individuals." Such a statement is as closely related to the practice of logic as the statement that "whales are fishes because they live in the sea" is to the practice of biology. Nor are "classes" in contemporary logic the sorts of necessarily eternal, necessarily one-attribute bearing pseudoentities this literature claims they are.

Second, the species = individuals literature, as well as much other evolutionary discourse, grossly misrepresents the history of philosophy. There is indeed a traditional "problem of universals," in which philosophers, beginning with Boethius, inquired about the reference of general predicates: when I say, for example, "This book is red," or "This computer program is user-friendly," I may speculate on what such a general predicate as red or user-friendly "stands for." Is there some real entity somehow somewhere to which such terms refer, or are they concepts in my mind, somehow obtained by abstraction from particular sense experiences, or are they just words, convenient for practical purposes but with no referent either in the world or in my mind.

Philosophers throughout the centuries have taken different views on this question. They were not, as is often alleged, uniformly "essentialist" in their answer to it. Nor—and this is the chief point here—has the problem of universals had much if anything to do with the biological problem of natural kinds. So much the worse for philosophers, one may say; they should be thinking about biology—but they were not. Even Aristotle, who was himself a great biologist as well as the ruling philosopher of many centuries, was not primarily interested in what we call species. As has recently been established, he uses the terms *eidos* and *genos* (which we render *species* and *genus*) entirely unsystematically and almost interchangeably in his biological works (Balme 1962; Pellegrin 1986).

At the same time, as is clear from our reliance on Hull (1980), we do indeed recognize that biological entities other than single organisms may be taken as individuals, and that it is often important to make this move; in fact, it is fundamental for our argument. Not only species, but monophyletic taxa in general, and, just as significantly, communities and ecosystems may and should be taken to be real, individual entities playing distinctive roles in the evolutionary process. Where possible, it is perhaps better to call all these "historical entities" rather than individuals, in order to avoid entanglement in that very bewildering literature.

At the same time, we may perhaps remark in passing (although the point is not essential to our argument) that there may also be reasonable and permissible senses in which species may be taken to be natural kinds, though not, of course, eternal or defined through reference to single, taggable properties. Indeed, all kinds of entities may exist and have members without being unchanging or eternal. I may belong to an organization, for example, that originates, develops, and goes extinct; so long as I and it exist, and I pay my dues, I am a member of it. That doesn't make it eternal or "essentialistic" any more than it makes me so. Such human artifacts are not natural kinds, but they *are* classes with members; indeed, they are paradigm cases of such entities. The point is that not every entity characterized by class membership is thereby eternal and unchanging. And so, just because they are classes with members, natural kinds, by analogy, need not be eternal, single-predicate tagged "classes" in the peculiar (and invidious) sense in which species = individuals adherents wish to use that term. Of course they are also historical entities, as are all organized groups, whether produced by nature or by human decision-making devices. The strong disjunction insisted on by Hull and others between classes as eternal entities and individuals as developing and noneternal has given rise to unfortunate and unnecessary misunderstandings in this area.

Be that as it may, by starting with Hull 1980, we hope to derive the benefit of the background debate he is moving from without entangling ourselves in its unfortunate details. And by moving in the direction of hierarchy theory we hope to show how Hull's distinction between replicators and interactors may be modified and put to use in a new and fruitful direction. (Hull does indeed mention hierarchy but does not, so far as we know, use the term in any systematic way.)

Perhaps, however, we should make explicit at this point the sense in which we intend to use the species concept in subsequent chapters. (For a more detailed account, see Eldredge 1989, 1992.)

The prevalent conception of species in the period of the evolutionary synthesis has been the biological species concept (BSC), which takes species to be "groups of actually or potentially interbreeding natural populations that are reproductively isolated from other such groups" (Mayr 1970:12). On this view, reproductive isolation accounts for the separation of organisms into the discontinuous collections that we call species. This concept, however, as both its proponents and its critics have acknowledged, does not hold true on an evolutionary time scale. In a macroevolutionary perspective, the attempt to demonstrate reproductive isolation becomes meaningless. Further, in Darwinian terms, despite real as well as apparent gaps in the record, the history of life does in a sense form a continuum. After all, in reproduction, one does have to go from one generation to the next. Moreover, the stress on species as reproductive systems seems to undercut the effort of systematists, pre- as well as post-Darwinian, to characterize species through the possession by their component organisms of a great many shared characters, by no means all of which have directly to do with their existence as reproductive communities. Human beings, for instance, not only mate with one another rather than with members of other species; they also have big brains, opposable thumbs, and so on. In short, in our terminology, it is economic as well as reproductive properties that mark off the organisms of one species from its congeners. In addition, the BSC (or the isolation concept, as it is now sometimes called) uses a negative character to mark off one taxon from another; but in the practice of many systematists, a negative characteristic is unsatisfactory as a classificatory tool.

Now in view of these, and other, oddities of the BSC, there have been numerous efforts to replace or modify it in favor of an "evolutionary" or "phylogenetic" concept that stresses the real existence of species over time. G. G. Simpson's (1961) evolutionary species concept, indeed, carried subjective overtones: species seemed in the last analysis to be artifacts of the investigator. Other more recent evolutionary concepts, however, like that of Wiley (1981), or the phylogenetic concepts of Cracraft (1989) or Mishler and Brandon (1987), or the cladistic concept of Ridley (1989) are attempting, much more emphatically than Simpson, to single out species as *real* evolutionary, in particular monophyletic, units. Much of the recent literature, in fact, draws also on the "species are individuals" thesis. The effort is to describe species as real, historically bounded entities existing throughout a given segment of evolutionary time.

Our usage in the present work follows this tendency, but with a difference dictated at least in part by the problem with which we shall be dealing. Given the difference between reproductive and economic adaptations that we put at the head of our analysis, we wish to focus on species as reproductive (genealogical) rather than economic entities. In this we concur with adherents of the BSC. At the same time we believe that both the negativity of the isolation concept and its necessary restriction to synchronic contexts can be overcome by the adoption of Paterson's Specific Mate Recognition System concept (SMRS; Paterson 1985; Eldredge 1989, 1992). Economic properties also play their roles in the existence and extinction of species but the nature of species, in our view, is genealogical. Species, we say, are "reproductive communities composed of organisms sharing a single fertilization system; they are distributed discontinuously; and their component organisms play analogous economic roles in different local ecosystems" (Eldredge 1989:132).

These matters will be dealt with in what follows. However, let us anticipate, and address if we can, some objections we know will be raised by systematists who would resist our focus on sexual reproduction as the necessary criterion for discriminating between species.

First, there are, of course, some purely asexual "species." These, we maintain, are in effect recognizably differing clones, brought into being primarily by mutations (Eldredge 1989:122, n. 8). More important, there are many organisms that differentiate economically as conspicuously as they do genealogically. In these cases systematists (especially botanical systematists) will want to distinguish species that in fact share an SMRS and that by our definition we would be unable to separate. In such cases, the SMRS becomes plesiomorphic: that is, it fails to distinguish groups based on derived characters (see fig. 1 in Eldredge 1992). We appear to have here two well-grounded versions of the species concept, each appropriate to its own task. Recent proponents of the phylogenetic species concept are dissatisfied with this form of pluralism (Mishler and Brandon 1987). Whatever the outcome of this debate, however, we are accepting one side of the controversy here: the emphasis on species as primarily reproductive rather than economic entities. At the same time we recognize that systematic interests may dictate, in other circumstances, a concept that differs from ours.

So much, then, for our proposed use of "species." Finally, we should note that, apart from the two-hierarchy conception we are relying on, there are other directions in the current evolutionary literature that lead to an expansion rather than a rejection of Darwinism, but in ways that appear, so far, rather different from the path we are taking. Whether a new "new synthesis" will bring all these investigations harmoniously together remains to be seen. The literature of developmental constraint, for example, adds to evolutionary biology a dimension it has always admitted lacking but is only beginning to fill in (Maynard Smith et al. 1985). Constraints may constitute limits in a relatively uninteresting sense: obviously, selection can operate only on what is already extant, that is, on what past selection has produced. But constraints may themselves be construed as part of the causes of evolutionary events, thus introducing within a basically adaptive view some nonadaptive principles: principles of self-organization, as Kauffman calls them (1985, 1992) or of structure (Goodwin 1988; Webster 1984).

A comprehensive evolutionary theory, and even a comprehensive theory of social systems, should perhaps, and will perhaps ultimately, embrace such complicating themes. Regrettably, however, it is not yet clear how all those modifications and expansions of traditional theory can be unified. Maybe development itself is hierarchical, maybe not; if it is, then a generalized hierarchy theory should take it into account. Maybe self-organizing principles enter into the evolutionary process at more than one level of the genealogical hierarchy, if not, indeed, in the ecological hierarchy also. Again, in that case, our speculations should eventually coalesce with the theories of such evolutionists as Kauffman (1992) and a few others.

It is beyond our hope or competence, however, to undertake so grand a synthesis here. If we can use recent work in hierarchy theory, and the enriched Darwinian perspective it permits, to provide a reasonable account of the origin and nature of social systems in animal life on this planet and in human life in particular, that will have to do as one contribution to the larger project of a newly developing evolutionary synthesis.

CHAPTER TWO

•••••••••

Darwinism
and Ultra-Darwinism

Although in the history of the sciences great names loom large in many areas, very few indeed have come to dominate whole disciplines as the name—and the thought—of Charles Darwin has come to dominate, whether more or less directly, the fundamental beliefs, and to a large extent the practice, of biology. True, not all biologists are practicing evolutionists, directly addressing evolutionary problems in their day-to-day work. Many ignore evolutionary questions and evolutionary theories for much of their working lives. But whatever their disciplines—from biochemistry to ethology, or from microbiology to paleontology—they accept the fact that Darwin (with others, but chiefly Darwin) established: the fact of evolution, or, as he called it, descent with modification. Hence, for example, the absurdity of so-called "creation science"; Kitcher's book on this topic is aptly called *Abusing Science* (1982; see Eldredge 1982; Futuyma 1983; Newell 1982). Fortunately, that is yet another question we need not go into here. Moreover, despite the many attacks in the literature, both past and present, on Darwin's explanation of descent with modification, that is, his theory of natural selection, most biologists who specialize in the study of evolution, it seems fair to say, still accept in outline Darwin's account of selection as the chief agency of evolutionary change.

Whole libraries have been produced on this topic: on the character of Darwin's achievement as a scientific discoverer and on the nature and scope of his influence in what appear to be recurrently waxing, waning, and again waxing phases. Since neither of us ranks as even a minor executive in what is affectionately called the Darwin industry, we shall not attempt here yet another analysis in any detail of either of these questions. At the same time, the enrichment of evolutionary ontology that we depend on for our approach to the problem of the origin of social systems in nature demands for its understanding some grasp of the Darwinian ontology we are hoping in some respects to solidify and in others to revise. In particular, we want to look at the overall conceptual structure of the Darwinian tradition, what we might call the physiognomy of Darwinism, in its relation to the Cartesian-Newtonian scientific revolution. Darwinism, we shall argue, contains some of the characteristic features of that earlier vision but also contains features that foreshadow an end, for the biological sciences at least, of the Cartesian-Newtonian perspective.

Accordingly, we shall look first, very sketchily, at the place of biological explanation in the tradition of early modern science; second, at the general character of Darwinian theory in its relation to early modern "mechanism"; third, at some of the fundamental metaphysical ambiguities of Darwinism associated with that historical situation. As to Darwinism today, to confirm its influence in current biological research, we will mention briefly a few illustrations of a routine or harmless application of Darwinian principles, speculative and empirical, before proceeding, fifth and finally, to consider the way in which, despite the acknowledged power of a Darwinian heuristic in many areas, a particular "thin" version of Darwinism that we call ultra-Darwinism dilutes and distorts the richer theory of Darwin himself.

Biological Explanation and the Scientific Revolution

If history as it was happening is hardly relivable, we nevertheless live, each of us, and each social group, within a history that is what it is in part through the way we perceive it. The practices initiated in Padua by Vesalius, and carried by his successors in a direct line

to Harvey, were surely, for the anatomists of those days, the major
events of their time. They saw, handled, and understood the fabric
of the body in a new way. And, of course, in any history of anatomy
and physiology their importance is recognized. Galileo, however,
working in Padua half a century after Vesalius and in the very time
when Harvey was a student there, was unconcerned with the explanation of living structure or function. Indeed, nature revealed herself to him in her true nature only when the living creature was
removed. The very qualities associated with life, and especially with
sentience, were barriers to understanding the pure mathematical
language of Galileo's world. It is that vision of dead nature that, in
our later view, constitutes the scientific revolution. God said, Let
Newton be! and all was light. For Newton set forth the universal
laws of motion (as distinct from such specific cases as animal locomotion), the law of gravitation holding for all bodies whatsoever,
whatever their size, shape or function. This was nature mechanized
and mathematicized—not that the two idealizations always marched
together (Westfall 1977). Still nature did consist, ideally, of hard,
solid, impenetrable particles, to be studied with mathematical rigor
at an instant, as they coursed uniformly through a true mathematical time.

There is no room for living things in such a world. Although the
Cartesian notion of a beast-machine was never universally accepted,
it was nevertheless symptomatic of its time. To conceive of living
bodies (our own included) as machines freed the mind to find its
own empire, including its mathematical grasp of physical laws in
their grand abstractness. How far acceptance of the *bête-machine*
really enabled experimenters to believe they were not inflicting pain
on their victims it seems impossible to say. But the ideal of a purely
mechanical world was certainly an influential one—one that in an
age of high tech still manifests its sway.

How in such a world, does one go about explaining "scientifically"
the immense variety, the myriad oddities, that seem to characterize
the living scene, the "nature" of natural history? The regnant Gali-
lean-Cartesian-Newtonian ontology could undertake such explanation only uneasily, if it cared to do so at all. And even though the
Darwinian revolution, two centuries later, helped us (or should have
helped us) place ourselves within the living world, it still moved
uneasily within the confines of that older, loosely speaking "mechanistic" tradition. We shall look at that situation; here the point is

just to indicate, as our starting place, the anomalous position of life in the *explanandum*—and the *explanans*—of the chief modern European cosmology.

A conspicuous feature of the new image of nature was its abolition of final causes. Natural processes, in the Aristotelian perspective, always possess a telos, a "that-for-the-sake-of-which." This is, in the case of animals, the production of the adult form (the male form in particular, according to Aristotle, which the female fails to attain). The starting point of a natural process bears an intrinsic relation to the particular end to which (other things being equal) it proceeds. Inquiry into ends, however, is inappropriate for a modern investigator of nature. There are, indeed, God's purposes, which it is beyond our limited powers to search out in any detail; but in examining the overall regularities of nature as God made it, we must restrict our search to the laws of motion, impact, and the like.

Roger (1963) points out that in the life sciences of the time the search for general mechanisms gave way to descriptions of particular mechanisms, bits of machinery in the wonderful engineering project by which God had made His creation. But in both the Cartesian and Newtonian traditions the search for universal mechanical laws was certainly predominant. Practitioners of Cartesian physics investigated the relation of the positions of bodies within a system of vortical motions. They carried through their master's resolution to ignore or deny other forms of change talked of by scholastic philosophers—such as birth and death (substantial change, in Aristotle's terms), growth, or change of quality—for all quality, in the newly conceived nature, came down to changes of geometrical relations. Only local motion remained, with no natural terminus in any that-for-the-sake-of-which. And Newtonian physics, for all its rejection of Cartesian models, retained this restriction of natural change to local motion—change in place.

Granted, not all anti-Aristotelian philosophers were so radical. Gassendi, as bitterly opposed to scholasticism as any one, had no objection, it seems, to final causes in the explanation of living things, although for him, as with so many of the leading innovators of the time, that was a relatively minor theme in a more generalizing vision of nature (see Bloch 1971). And there were important figures in the eighteenth century also, notably Maupertuis (following Leibniz), who tried to reintroduce or to retain final causes even in physical explanation, pointing to the law of least action as an instance of final causality in nature (see, e.g., Buchdahl 1969). At the same time, it may still be said that in general the move to abolish final causes from natural explanation formed a predominant attitude in early modern thought, and the "efficient cause," divided from teleology, was readily translated into a concept of local motion, or change of place, as the paradigm for all natural events. In the scholastic tradition, natural events or entities, being finite, have a proper end toward which they move as well as a characteristic moving cause that triggers their beginning. In modern thought half of this pair of "causes" was cut away; cause became just the first item in a sequence, which set off the rest: the set of necessary and sufficient conditions.

Further, the other traditional pair of causes, "formal" and "material," the question of what a thing is or what its function (its *ergon*) is, and the question of what it is made of, were split apart and both transformed. Instead of the proximate matter bearing on a given form, or style, of organism—as flesh and bone are matter for an arm, for instance—"matter" itself was investigated simply by being analyzed into its parts. The scholastic phrases *substantial form, quiddity, real quality,* and so on appeared to be, and indeed were, useless jargon masquerading as explanation. They did no work—the School, Descartes insisted, had never solved a single problem—while the treatment of physical questions in terms of the pure spatial relations of the parts of bodies could produce a whole science of physics that would solve problems one after another in the proper order. In Descartes's view, indeed, matter *was* spatial relation. Except in the unique case of the human mind-body relation, the concept of form had become, for Descartes, burdensome and unnecessary.

It is an oddity of history that some of the most popular Cartesians, notably Rohault (1671b), readopted form-matter vocabulary and insisted that Cartesian shapes were Aristotelian forms and Cartesian extension (spreadoutness) was Aristotelian matter. There is so little resemblance in meaning and use between the two pairs of principles that one can only suppose it still seemed politic in late-seventeenth-century France to attempt such an implausible rapprochement. Fundamentally, however, Descartes was surely right about his own radical opposition to scholastic physics. Instead of many differently qualified beings, each with its own "substantial form," we had one extended universe, made by God in one creation. Whether its regular behavior developed out of an initial chaos or was plain from the beginning makes no fundamental difference. Whether we follow the early project of *Le Monde* or the somewhat altered version of the *Principles,* it is clear that God made an infinite, or better, indefinite plenum that follows the simple laws He, in His constancy, prescribed for it (Descartes [1632] 1974, [1644] 1973c). *We* come to know the operations of His universe by analyzing the relative shapes, sizes, and positions of the bodies moving, relative to one another, within that vast, spread-out nature. Form, if it survives at all, seems to become shape, which it had sometimes been in Aristotle, but by no means always or essentially. It is material causality, how the parts impinge on one another so as to produce motion, that are basic to physical inquiry and explanation.

In general, on Cartesian grounds, all biological phenomena must be explained mechanically or (to bend a little that all-too-flexible concept) in terms of machinery—that is, through an account of the way in which the parts of automata are arranged so as to perform (or to seem to perform) some operation. The analogy with popular automata of his time, which Descartes used frequently, was carried further by his disciple Rohault (1671a). If a clock can be made to tell the time, he argued, using only ten basic parts, how much more can the animals designed by God accomplish with their many carefully arranged parts and organs? They are exceedingly complex machines, but still machines, explicable, ultimately, in terms of the interaction of their parts according to the universal laws of motion decreed by God at the Creation. Such intricate interactive systems may at present or even forever elude our finite powers of analysis, but on principle, as with any machine, there is nothing there to resist their dissection into their constituent parts, which, in turn, by following the simple laws of motion, make up their seeming wholes.

Cartesian nature is a plenum, characterized by vortical motion. Analysis into parts, in such a nature, goes on indefinitely: there are no least parts, only very little ones. Descartes did divide his matter into three elements that are relatively dense or rare—and in particular his "subtle matter" has to rush into the pores of rarefied bodies, which seem to have "empty" interstices. In such a situation, a distinction between relative and absolute motion is difficult to make. Indeed, the force to initiate motion has to be applied from outside, by divine decree.

The alternative to a Cartesian plenum, of course, which proved more successful, was an atomistic physics such as that pursued, or at least suggested, by Boyle and Newton. If, as Newton alleged, God probably formed matter out of hard, solid, impenetrable particles, then the question "What is it made of?" is definitively liberated from its alliance with forms, quiddities, and such. We can explain the behavior of bodies in terms of the behavior of their least parts, and all mystery (within God's creation, of course) is eradicated. Granted, such atomistic explanation depends on contact between the parts of bodies. There must be no action at a distance. Gravity, therefore, seems an anomaly. Newton's disciple Samuel Clarke considered that it must be the effect of some immaterial property added to matter by God (Buchdahl 1969:386).

Whether in terms of French vortices or British particles, however, the elaborate and seemingly self-perpetuating rhetoric of the scholastics was defeated at last. Explanation was open-ended: inquiring into the moving cause of a limited, goal-directed process was replaced by the endless possibility of billiard ball causality. There were no more "natural places" sought by traditional elements, nor, indeed, was there any differentiation in the natures of particular bodies, nor, correspondingly, in the differentiation of separate sciences. The vision of natural laws that would explain all phenomena everywhere offered a dazzling prospect.

Ideally, moreover, for the new science, explanation was mathematical. For Aristotelians, mathematics was one science among others, characterized, indeed, by a lack of full reality in its subject matter. Mathematics studied either discrete or continuous quantity; but bodies, full-blooded bodies, are much more than quantities. Aristotle reproached Democritus for attempting to explain all natural properties in terms of size and shape alone. Using, as he often did, an artifactual analogy, he pointed out that shape alone could not make a threshold: it has to be solid, to support the tread of those who enter the house (see, e.g., Grene 1964). The new corpuscular philosophy, as it came to be called, reverted to this old Democritean habit. Of course, the new thinkers went much further than Democritus in the mathematization of nature. Descartes, although no atomist, had boasted that his physics was entirely geometrical. And although it is hard to find much mathematics in his *Principles*, one can see, in a way, what he meant. His account of refraction in the *Dioptrics*, for example, relying on the analogy of a blind man feeling his way with a stick in order to explain vision or on the case of a tennis player striking his ball, makes spatial relations fundamental to physical events in a way that is at bottom "geometrical" (Descartes [1637] 1973a). And, of course, it was Newton's mathematical genius, not experimental skill, that produced his grand synthesis. Looking back at the traditional form-matter relation, one could perhaps suggest that while moving cause had declared its independence of telos and material cause had been let loose to be investigated by analysis of bodies into their parts, form had been replaced at first by shape but ultimately by formula.

The success of the new mechanical philosophy suggested to many thinkers a parallel in the laws of nature that govern human life. For instance, Richard Cumberland (1672) considered that the laws of universal benevolence and all their consequences could be "reduced to a proper similarity with the propositions of universal mathematics, concerning the effects of mathematical computation, through which all quantities are brought together with one another." But where did this leave living things other than ourselves? On the grand scale, nowhere in particular. As Löwith (1967) and others have argued, the task of modern ontology was to construct a world out of three entities (or kinds of entities): God, man (or mind), and one-level, spread-out nature, whether continuous or particulate. Living things are an oddity, and some odd people may show an interest in them, but they have no central place in our conception of the world around us. Microscopists might amuse themselves with their new worlds within worlds; Leeuwenhoek might become a foreign member of the Royal Society; the debate between ovists and animalculists might rage. But that did not alter the basic Cartesian-Newtonian cosmology and the problems it raised if one wanted to ask biological questions and understand biological phenomena in its terms.

Characteristic of the reigning modern vision, in particular, is the distinction between primary and secondary qualities. Locke, following Boyle, made the terminology current, but the distinction is posed in other terms by Galileo, Descartes, and that unpopular but prophetic writer Thomas Hobbes (Locke [1690] 1975; cf. Buchdahl 1969). It goes, of course, with a geometrizing and/or atomizing conception of what there really is "out there"—bodies in motion. These may be described as purely extended, or, in addition, impenetrable, as infinitely divisible or composed of least parts, whose fundamental properties sum up to produce the properties of larger bodies. In any version, what is plain is that while our perceptions of size, motion, and other "primary qualities" (if there are others) bear some relation to the actual characters of real physical things, our perceptions of color, taste, odor, and sound are in themselves subjective. They are founded, in the views of Locke and Boyle, on "secondary qualities" of bodies, which are in turn effects of their primary qualities unknown to us in any detail; but they bear no qualitative relation to the character of the real bodies of nature or their least parts. They would all, said Galileo, be "mere names if the living creature were removed" (Galileo 1957:272).

The most famous recent version of this division is Eddington's tale of the two tables, one composed of electrons buzzing about in empty space, the real, physicist's table, and the other a hard brown object one can rest one's plate or paper on, a mere subjective construct, a fictional substitute for the real world. Until recently even the psychology of perception has been founded on a crude version of the primary/secondary distinction (Reed 1988). Yet living things, at least animals, and indirectly also many plants, in their ongoing relations to their environments, depend for their very existence on colors, odors, tastes, noises, or melodies. Mate recognition, prey or predator recognition, let alone phenomena like mimicry, depend on sensory cues that inform the organism in question of the character of other objects and events around it. But there are no environments in the Newtonian universe, only tasteless, soundless, odorless, colorless motions through empty space.

This is not to deny the achievements of Linnaeus and other early modern taxonomists, or of such biologists as Grew, Hales, and so on. Most of them, like John Ray, could work happily at observing the diversity and ingenuity of God's creation. But to the great sweep of natural philosophy they contributed little. There was occasional quantitative work in biology, too, of course, like that of Hales. But on the whole it must be said that the universal laws of nature passed living nature by.

In this respect we may take Kant's critical philosophy as symptomatic of the period that stretches from Descartes to Darwin. In an early precritical essay, Kant proclaimed: "Give me matter, I will build a world of it!" That was for the physical world alone, however. Could one make the same claim, he went on to ask, for "the least of plants or insects"?

> Is one in a position to say, *Give me matter, I will show you how a caterpillar can be produced?* Is one not held up here at the very first step out of ignorance of the true inner constitution of the object and the intricacy of the manifold it contains? It should not be surprising, therefore, if I venture to say that the formation of all the heavenly bodies, the cause of their movements, in short, the origin of the whole present constitution of the cosmos will become intelligible before the produc-

tion of a single plant or caterpillar becomes plainly and completely known on mechanical grounds. (Kant 1968:230)[1]

The same sense of the difficulty of assimilating organic nature to the world studied by science underlies Kant's Third Critique, where the methods of biology, and, in particular, the union of teleology with mechanism, form one of the two chief subject matters of the work. Kant's philosophical treatment of biology influenced deeply the work of German biologists for the first half of the nineteenth century (Lenoir 1989), yet his approach to this seemingly less than tractable domain remained uneasy. Biology somehow evades the rigors of strict science.

There would never be, Kant predicted, a Newton of a blade of grass, that is, a law giver for biology such as Newton had been for the inanimate cosmos. True, after Kant's time there were various candidates for that honor: some claimed Cuvier, and it is reported that Owen offered himself for the position. How Darwin proved Kant mistaken, and how Darwin's achievement related to the Cartesian-Newtonian tradition, is what we have next to consider.

Darwin and the Heritage of Newton

To many evolutionists, Darwin is the St. George of biology, the writer-naturalist-theorist who overcame at one blow the denial of life characteristic of the Cartesian-Newtonian vision. He not only placed *us* where we belong, in the midst of nature, but he also brought a whole dead nature back to life. Thus Dobzhansky (1968) published an essay in which he contrasted "Darwinian" with "Cartesian" approaches to biology. Yet in some fundamental respects Darwin still remained firmly within the older tradition; his "revolution" was not as abruptly revolutionary as some have claimed.

In methodology, indeed, there is a direct historical line linking Darwin back from Herschel and Lyell to Newton's "first rule of philosophizing": Rule I. We are to admit no more causes of natural things, than such as are both true and sufficient to explain their appearances (Newton 1962, 2:399).

M. J. S. Hodge's (1977) careful analysis of Darwin's argument in the *Sketch,* the "Big Book," and the *Origin* itself makes it clear how closely he had followed Herschel's rendering of this maxim. The *vera causa principle,* Hodge argues, must contain two parts, the first double: the alleged cause must be shown on independent grounds to be both *true* or *real* and *competent* to produce the results claimed for it; and it must be reasonably held to be *responsible* for those consequences. It is true that in Newton's formulation "competence" and "responsibility" are not distinguished, and in many evolutionary explanations it is difficult to distinguish them—hence the "just-so story" complaint about many selectionist explanations of biological phenomena (Gould and Lewontin 1979). If an evolutionary account is consistent with the theory of natural selection, it is often held to confirm that theory. However, the second half of the *vera causa* question is not "could it have happened that way?" but "did it *in fact* happen that way?" That Darwin himself was sensitive to this distinction is clear in the conclusion of part 1 of the *Sketch* of 1842. He writes:

> If variation be admitted to occur occasionally in some wild animals, and how can we doubt it, when we see thousands of organisms, for whatever use taken by man, do vary. If we admit such variations tend to be hereditary, and how can we doubt it when we remember resemblances of features and character—disease and monstrosities inherited and endless races produced (1200 cabbages). If we admit selection is steadily at work, and who will doubt it, when he considers amount of food on an average fixed and reproductive powers act in geometrical ratio. If we admit that external conditions vary, as all geology proclaims they have done and are now doing—then, if now law of nature be opposed, there must occasionally be formed races, differing from the parent races. . . . Are not all the most varied species, the oldest domesticated: who would think that horses or corn could be produced? Take *Dahlia* and potato, who will pretend in 5000 years that great changes might not be effected; perfectly adapted to conditions and then again brought into varying conditions. Think what has been done in few last years, look at pigeons, and cattle. With the amount of food man can produce he may have arrived at limit of fatness or size, or thickness of wool, but these are the most trivial points, but even in these I conclude it is impossible to say we know the limit of variation. And therefore with the selecting power of nature, infinitely wise compared to those of man, I conclude that it is impossible to say we know the limit of races, which would be true to their kind; if of different constitutions would probably be infertile one with another, and which might be adapted in the most singular and admirable manner, according to their wants, to external nature and to other surrounding organisms—such races would be species. (Darwin 1958a:57-58)

Thus the existence of selection and its competence to produce new species may be established, or made plausible, perhaps overwhelmingly so. But then there is one more decisive concluding sentence: "But is there any evidence that species have been thus produced, this is a question *wholly independent* of all previous points, and which on examination of the kingdom of nature we ought to answer one way or another" (italics ours; Darwin 1958a:58).

Following through the twofold (or in one sense threefold) articulation of the *vera causa principle,* Hodge finds the *Origin* organized as follows: chapters 1 through 8 deal with the first half of the principle—chapters 1 to 3 establish the existence of selection, chapters 4 and 5 make a case for its causal competence, and chapters 6 to 8 consider the difficulties entailed by such a claim, notably with respect to instinct and to hybridism—while chapters 9 to 13 consider the question of causal responsibility. It does seem odd that the third and most positive section, which argues for the thesis that natural selection has in fact been the cause of the origin of species, should begin with a chapter on geological *difficulties,* but on the whole Hodge's argument is certainly convincing. Indeed, on the evidence of the *Sketch* alone there would be no denying Darwin's scrupulous and self-conscious use of the traditional Newtonian-Herschelian *vera causa principle* as the foundation for his theory.

Situating Darwin's methodology within this tradition in a general way, however, says little if anything about his metaphysic. Anybody can search carefully for true causes and, depending on his/her ontology, discover them in different places—as we shall see when we consider the ontology of ultra-Darwinism. What we want to claim here, further and more substantively (though we admit in very global and speculative terms) is that our Newton of a blade of grass still retained elements of what had once been called "the new mechanical philosophy."

"Mechanism" has always been an ambiguous term. In its strict use in the seventeenth century it referred to the art of making engines, weight-lifting devices, and so on. When Huygens asked Descartes for "five pages of mechanism," Descartes sent him instructions for building such gadgets. And when a critic complained that the essays published in 1637 were "too coarse and mechanical," Descartes replied proudly that mechanics had always been the reasonable part of physics, uncorrupted by scholastic verbiage, because if people failed in a "mechanical" enterprise, they would lose their investment—so they couldn't afford to clutter up this practical discipline with the high-sounding terminology that had led most of physical inquiry astray (Grene 1989).

"Mechanism" in this sense—in which we search for *a* mechanism to embody a process-has always lingered in the background of mechanistic science, and especially in biology, where we look for the function of an organ or organelle, which we then use for an explanatory purpose, much as we might explain to a novice the function of a carburetor or a catalytic converter. Whether or not organisms are held to be machines in the Cartesian sense, their organs do seem to be "mechanisms" put to use by them in a regular and reasonable way. If, as the unlikely story goes, Newton was inspired to his greatest discovery by watching an apple fall, it is perhaps just as fateful that when his mother sent him to watch the sheep, he absentmindedly set to work designing a new kind of mill, driving his exasperated parent to bewail her son's incompetence (Westfall 1980). Machines for grinding grain or machines for keeping the heavens and earth in order: it's machinery that matters.

At the same time, of course, when we think of the "new mechanism," it is not so much engineering we think of as a certain style of causal explanation. Boyle, later in the century, explained that he was using *mechanical* not in the strict sense, that is, to refer to the art of designing machines, but in a broader one, and it is that sense that seems to be primary in the new mechanical philosophy. It is a question of bodies in motion, following certain simple and universal laws, so that from one place they are propelled to another. The model works best, perhaps, if one takes up the ancient atomist view: there are little bodies moving in empty space, some of which, or many of which, cohere by some kind of simple physical entanglement to form the bigger bodies we seem to see around us. And it is the physical impulsion of such bodies or collections of bodies that constitutes all the phenomena of nature.

What is a causal explanation in this kind of universe? Causality used to be linked with power, with the greater reality or perfection of one thing or disposition or property, such that it could effect, out of its own overflowing actuality, the alteration or coming into existence of some lesser being. Thus the doctor's art, when effective, could cause health, or the statesman's, peace, and the superior reality of the (male) parent could, with the help of female matter, produce an offspring. All that nonsense is gone. There are just plain bodies spread out in space, and as one bumps into another, a succession of events occurs that, taken together, constitute "nature." Cause and effect are a linear sequence of such collisions and consequent repulsions. This is mechanistic explanation, where the position of one collection of particles at one time plus the laws of motion suffices to predict the position of those particles at the next time. Its culmination, of course, is in Laplace's omniscient observer. This is the simple when-then, one-level determinism that characterized classical physics until this century.

At the same time, however, the superiority of the new physics was said to reside in its mathematical character: Descartes boasted that all his physics was geometrical, and it was certainly Newton's brilliant quantifications—his geometrical proof of Kepler's laws, the quantitative expression of gravity, the invention of the calculus (the "method of fluxions")—that secured him a unique place in the new science. That mathematical motif, however, is clearer in twentieth-century Darwinism than in the work of Darwin himself, unless we count the simple Malthusian "law" that seems to have triggered his discovery as "mathematicization." On the whole, it appears that Darwin was not only himself not a mathematical thinker; he was inclined to suspect the reasoning of mathematicians (and mathematically tutored physicists) in matters pertaining to biology. In twentieth century Darwinism, indeed, there is no doubt that the use of mathematical procedures has added dramatically to the prestige, as well as the experimental efficacy, of the theory (Provine 1971). We shall return to that point briefly below. But in thinking of Darwin himself in the context of the "mechanistic" tradition, it is not so much with the hope of quantification as with the ambiguity of the concept of mechanism itself that we have to deal.

Mechanism as machinery is clearly at the center of Darwin's thinking. Indeed, if we go back to his interest in Paley's *Natural Theology* (1802), one of the few texts he admits to having enjoyed reading as an undergraduate, we can see that the concept of organisms as adaptation machines lies at the heart of his theory. In living beings we find an amazing host of *devices for*—feeding, breeding, moving, hiding—almost everything of interest in organisms seems to be cunningly devised to achieve some "end." In Paley's view, of course, all this cunning had to be attributed to an Infinite, All-Wise, All-Good Maker of all things; but if in the light of the infinite variety of nature we could explain how all this diversity had come about by natural means, that would be, in line with the Newtonian hope of finding uniform causes for a wide range of phenomena, a more satisfying way to explain at one gasp an otherwise bewildering diversity.

The point here, however, is not yet the means of explanation Darwin was to discover, but the thorough mechanistic character of the phenomena to be explained: "mechanistic" in the original, engineering sense of the term. An organism is a kind of embodied ways and means committee. It is a collection of gadgets for getting a collection of jobs done: in short, an adaptation machine.

This aspect of living nature was constantly to fascinate Darwin: witness the 1862 work on orchids, republished in 1877 with the slightly abbreviated title: *The Various Contrivances by Which Orchids are Fertilised by Insects* (Darwin 1862). "Contrivances," that's the name of the game. Shaw said Darwin threw Paley's watch into the ocean. However, as has often been noted, most recently by implication, in Dawkins' newest "Darwinian" credo, *The Blind Watchmaker* (1986), it was the foresighted watch*maker*, not the watch, that Darwin dispensed with (in Dawkins' account, perhaps not even so much the maker as just the foresight!). Darwinian nature is chock full of watches, machinery that gets things done, only it doesn't take a supernatural Maker to make them.

In other words, Darwinian biology is profoundly, some would say incurably, functional biology. If there are inconsequential, useless characters thrown up by evolution, they are free riders on the mechanisms produced by natural selection, which have to be of some *use* to their possessor, and particularly of some use in facilitating differential reproduction, if they are to be "noticed" by that "agency." One of the most influential Darwinian treatises of the last twenty years, G. C. Williams's *Adaptation and Natural Selection* (1966), while perhaps best known for its vigorous defense of genic selectionism, puts that modish message scrupulously into the correct Darwinian context. Only analogues of engineering devices, Williams insists, are to be explained as adaptations, and therefore as produced by natural selection. A flying fish returns to the water by gravity; that is of no interest from the point of view of a theory of selection. But how it gets itself up there is another—and a good Darwinian—story.

What Darwinism investigates, then, is a congeries of mechanisms in the sense of engineering devices; that they have not been engineered by an Intelligent Designer in no way contradicts this fact. Let us put it even more broadly into the context of the Cartesian-Newtonian tradition. Earlier views of nature, we have seen, had worked in terms of the pair of analytical concepts, form and matter. Cartesianism as well as its more successful Newtonian descendant subordinated the study of form to that of a liberated, so to speak, free-floating matter. In biology, however, this means that morphology, the study of form as such, is demoted to second place and that the way the parts of living bodies are put together—in effect, their functions—takes priority over their form. Indeed, many opponents of Darwinian evolution both in the nineteenth century and today have objected to this reversal: they decry Darwin's theory because it is functional, not structural (Webster and Goodwin 1982; O'Grady 1984). This seems surprising, since function does seem to be essential to life: when living things stop functioning they no longer exist. But what we want to stress here is simply the idea that the neglect of an emphasis on form, with the concomitant stress on the priority of function, concurs well with the spirit of Newtonian natural philosophy.

However, if it was machinery (engineering devices) that Darwin, like Paley, saw everywhere in nature, it was through mechanistic explanation in the causal sense that he hoped to account for their origin and their maintenance. The seeming purposiveness of living structures and behaviors is emphatically, in the spirit of Newtonian-Kantian science, to be accounted for as purposiveness without an intrinsic, specifiable purpose. But such an account would no longer consist, as it seems to have done for Kant, of some arm-waving in the direction of God's goodness. It would be accounted for in plain, naturalistic terms, by pointing to the vast range of slight variation among various natural beings, whether that variety be produced spontaneously or by human effort, and to the differential fitness of such varieties for coping with the conditions of life in which they happen to find themselves, noting also the patent fact that offspring resemble parents, so that such differences in viability may well be passed on from one generation to another.

Since he had read Herschel and Humboldt as an undergraduate, Darwin had been filled, he would recall, "with a burning zeal to add even the most humble contribution to the noble structure of Natural Science" (Darwin 1958b:67). What a glorious contribution this new vision of nature would provide! It would substitute causality in accordance with uniform natural laws for arbitrary, multifarious, unaccountable acts of creation. It must have been indeed "a great blow and discouragement" when he heard that Herschel himself had called his book "the law of higgledy-piggledy" (Darwin 1882:2:26). Surely, it was the very opposite: he had spent twenty years marshaling data to support his new conception, in the true spirit of induction, whether interpreted in Herschel's sense or in the different spirit of Whewell. Indeed, he was patently following not only Newton's first rule, searching for *verae causae*, but the second, third, and fourth rules as well (not that he studied the *Principia*, but the spirit of the rules certainly prevails). Newton had said, in effect:

> Rule II. . . . To the same natural effects we must, as far as possible, assign the same causes.
> Rule III. The qualities of bodies, which admit neither intensification nor remission of degrees, and which are found to belong to all bodies without reach of our experiments [like the potato or the *Dahlia* of the *Sketch!*] are to be esteemed the universal qualities of all bodies whatsoever. (Newton 1962, 2:399)

And surely that is just what Darwin had done! He had shown how (useful) variation can be propagated, through cause and effect interactions of living beings with factors in their environments and through the inheritance of these slightly varying useful traits: ordinary when-then causality between organisms and the factors affecting them, like relative hardiness of a plant in a dry climate or relative ability to run fast when pursued by a predator, plus the ordinary sequential causality of parent-offspring likeness. Out of these perfectly plain causal relations, extrapolated to the whole of living nature, he had built a uniform, simply deterministic vision of a regularly changing world of living beings, in which varieties (incipient species) would naturally proliferate and equally naturally, in changing circumstances, go extinct.

Furthermore, even Newton's fourth rule seems to be exemplified in Darwin's repeated injunctions that his readers should not discard his theory because of difficulties entailed by it. Newton had declared:

> Rule IV. In experimental philosophy we are to look upon propositions inferred by general induction from phenomena as accurately or very nearly true, notwithstanding any contrary hypotheses that may be imagined, till such time as other phenomena occur, by which they either be made more accurate, or liable to exceptions. (Newton 1962, 2:400).

In other words, if you have a straightforward, mechanistic, when-then explanation that will account for a wide variety of phenomena, in the terms of rules 1 to 3, don't abandon it hastily unless equally weighty evidence provides an improvement or alternative—and the alternative in this case was no scientific account at all, but the doctrine of special creation. Thus Darwin's theory conforms to the spirit of the Newtonian tradition in both senses of the emphasis on mechanism: it finds in the natural world mechanisms that it proceeds to explain by means of mechanistic causal laws.

Tensions in Darwinian Explanation

At the same time, Darwin's mechanical explanation of the origin of biological mechanisms carries with it a number of further ambiguities, or perhaps tensions, that contribute both to its explanatory power and to its recurrent vulnerability to attack. We mention three.

First, although Darwinian explanation is deterministic (with a stochastic element built in, indeed, but still of the classical when-then style), it does contain what appears to be a conspicuous teleological component. The theory of natural selections holds, after all, only for adaptations, states of being adapted, adaptednesses; and these are means-end relations—it is not too much to say, Darwin concedes, that the hand is "meant for" grasping. Teeth are for chewing, wings for flying, legs for running. And so on. But in a truly nonteleological universe, nothing is *for* anything. We can say as loudly and as often as we like that it is only the *explanandum* of the theory that is "purposefully" organized and that the *explanans* is nonteleological—but there those ends, goals, "purposes" are. It is those that so concern Darwin and Darwinians and that they are constantly discussing. Do away with adaptation, and you do away with Darwinism. Worse yet, it has seemed to many defenders and detractors of Darwin's view that despite all disclaimers, all insistence that evolution is "bushy," not ladderlike, opportunistic, not geared to a given goal—despite that, there do seem to be overall goals to the evolutionary process itself: it seems to make progress, to go from "lower" to "higher" in a way that to a hard-nosed determinist is suspect indeed.

Darwin sometimes claimed, sometimes denied, progress in evolution, and so it has been for twentieth-century Darwinians as well. But the language of a kind of forward movement is hard to avoid. From monocellular to multicellular forms, from one layer of tissue to three, from "in"vertebrates to vertebrates, from small-brained to large-brained creatures: all these moves ahead seem to be there, broadly speaking, in the evolutionary record. Both the axiom of adaptivity—the fact that selection works only for useful traits—and the lure (or the snare) of evolutionary progress seem to infect this offshoot of classical determinism with a lingering but essential teleological component.

Second, there is the puzzling but insuperable tension in Darwinian thought between chance and necessity. Evolution can occur only if the needed variations to permit change at a given place and time happen to be available, so ultimately it is "all a matter of chance." So it has seemed to many of Darwin's critics: hence perhaps Herschel's "law of higgledy-piggledy." And even Darwinians, for example Dobzhansky (1937) in the earlier (not the later) editions of his classic *Genetics and the Origin of Species,* have sometimes appeared to welcome chance as a kind of ultimate in evolutionary explanation (cf. Gould 1983). When you get to the variations that just *were* available, that's rock bottom; one can dig no further. On the other hand, of course, selection necessitates: indirectly, through differential reproduction, it bends, and so builds, lineages of organisms in more viable and survival-prone directions. To the nonevolutionist, chance and necessity appear to be contradictories: what is by chance is surely nonnecessary, and what is necessitated is no matter of chance. Yet, as Monod's (1970) best-seller title shows, it is exactly the odd combination of these two incompatible "forces" that produces evolution in the Darwinian mold.

This global tension in Darwinian explanation may be illustrated more specifically by pointing to the tension between the two components of evolution, variation and selection. Variation, the stuff on which selection works, occurs (relative to the needs of the organism in its particular environment) by "chance." As the story is usually told, that is the first step in the two-step evolutionary process. Selection, the molding, necessitating aspect of evolution, acts to preserve, from one generation to the next, the better-adapted of alternative available variations. So it narrows the formerly available range of variations and restricts the very material needed for its own future effectiveness—more actively than Shakespeare's aging lover, consuming that which it was nourish'd by (Sonnet 73). Enthusiasts of the hardened-synthesis era sometimes forgot this tension. Thus de Beer (1958) pointed to the highly specialized beak formation of the huia bird as a beautiful example of the power of selection, but it also looks like a beautiful example of the decline of variability—if its bearer is extinct, no wonder! The more variation, the more material for selection to work on; but the more selection, the less variation, and so the less material for selection to work on. Variation by itself goes nowhere; selection by itself is self-defeating. The tension between the two "major factors" of evolution, therefore, is ineradicable.

Finally, there is a third pair of contraries, less conspicuous, but nevertheless involved in the Darwinian vision. This is the contrast between parts and wholes, or what we might oppose as items and contexts. The unique foundation of Darwin's achievement, as we all know—as Mayr and others have eloquently told us (e.g., Mayr 1982; Hull 1965)—was the turn from typology or essentialism to *population thinking*. What Darwin realized, thus transforming the very foundations of biology, was that it is not after all the type, the universal form of a species, that matters, but the slight differences between conspecifics that render some better able than others to succeed in the battle of life, and hence to leave behind slightly superior others resembling them in this fateful, if slight, superiority. It is not fixed, eternal universals but very minutely differing particulars the biologist should look for. Aggregates of these, *populations*, not archetypes, is what nature consists of, and it is these aggregates, over the generations, that change, little by little, to produce the multifarious crowds of living things, from barnacles to Britons, that now inhabit this teeming globe. It is the varying traits exhibited in particular barnacles or Britons, orchids or oysters, that have summed up and will sum up to the kaleidoscope of life's history. Although Darwin knew nothing of genes, attention to least parts, or at any rate to the small traits that serve as "parts" in the makeup of the organism, furnished the observational basis for his method. And particularistic thinking has ever since been a conspicuous attribute of Darwinian theorizing. Adaptation machines, like other machines, are aggregates of parts.

At the same time, however, natural selection envisages a history not only of the constituent parts of such machines but of the machines themselves and, indirectly but essentially, of the conditions to which they are or become, or fail to become, adapted. What the *Origin* tells us of is the "struggle for life", the comparative success or failure of particular beings, given their particular circumstances, to survive and leave descendants.

And that is in two ways a story of wholes rather than of parts. Even if, in twentieth-century style, we think of evolution in terms of differential gene frequencies, we have to recognize that it is organisms by which genes are carried and it is the fate of organisms—of whole plants or animals—that, as evolutionists, we are chiefly studying. Whether or no the story of color polyphenism in *Cepea nemoralis* is true or false (Cain and Sheppard 1950; cf. Lamotte 1959), it is a story about snails and thrushes and thrushes' anvils and meadows in summer and winter, not, directly, about genes. Nobody can think about selection, theorize about selection, draw conclusions about selection, without thinking on the one hand about whole organisms and on the other about what Darwin called "conditions of life": environments. So over against the partitionist perspective, the concentration on minute particulars, characteristic of Darwin's starting point, there is the stress on those living beings on whom such small particular differences are bestowing some slight but crucial superiority: in our terms, on phenotypes as against genes or genotypes. And beyond the whole organism in each case, there is, what is just as crucial for this whole conception of life's history, or of life as history: there is the environment to which the organism in question must adapt itself. There are matters of climate, soil, other organisms, including conspecifics, predators, prey—all these circumstances are essential to the fate of every individual. So there is a whole, what we would call an ecosystem, beyond the whole organism, and twice beyond the minute particulars from which our story starts, on which the very possibility of selection depends. Ecological as well as organismic thinking belongs to the very core of Darwinian biology.

The Influence of Darwinian Methodology

Determinism/teleology; chance/necessity; part/whole: how can a theory so riven with contradictory themes have exercised, and still exercise, so powerful an influence over so wide and varied a community of scientists—observers, experimenters and theorists alike?

In part, as we noted in passing, the introduction of mathematical models into evolutionary biology has given its constituent disciplines an aura of "real" science that they had previously lacked. Beginning students in genetics learn to work with the Hardy-Weinberg formula. In their applications of genetics to evolution, both Fisher and Wright relied heavily on mathematical formulations; indeed, as Provine has shown, their theoretical disagreement may well have been grounded in the difference in their mathematical approaches (1986). The school of MacArthur was founded on his introduction of novel mathematical models in ecology. And, in general, increasingly extensive and increasingly sophisticated applications of mathematical techniques continue to characterize much biological work.

But there is a more fundamental reason why so much evolutionary research continues to be dominated by the Darwinian perspective. Scientific theories are not algorithms. Even theories in physics, highly mathematical though they are in form, have to be understood in order to do their explanatory and predictive work. Darwin's theory, for all its tensions—indeed, in part because of those tensions—provided, and still provides, a guiding pattern for the study of an impressive range of biological phenomena. The concept of natural selection offers a pattern for problem solving, or better, for problem-raising and problem solving, that can be applied in empirical research and in biological theory in ways that make sense of hosts of otherwise diverse and incoherent phenomena (Brandon 1981, 1990; Kitcher 1985a). And whatever one's philosophy of science, it has always been understood, one way or another, that that is what theories are meant to do. They make the world tidy, so that we can see and understand its regularities. Neglecting for the moment underlying ontological implications and taking natural selection in its pristine form, as the consequence of slight phenotypic variation combined with differential fitness and the inheritance of differential fitness—taking natural selection in this most general Darwinian sense, we can see how purely much contemporary biological work still conforms both to the letter and the spirit of Darwin's original view.

It would be easy to multiply almost indefinitely examples of this point; anyone who wants a good sample of relatively straightforward applications of Darwinian theory in contemporary research should consult the list in chapter 5 of Endler's *Natural Selection in the Wild* (Endler 1986, table 5.1). But, of course, there are whole fields, such as behavioral ecology, ethology, behavioral genetics, and so on to choose from. At the Darwin centenary, for example, J. L. Harper, a plant ecologist, while pointing out that ecology had long failed to follow Darwin's methods, starts from a passage in the *Origin* and proceeds to report work of his own and others following so closely from this text that he can describe the work in question as "quintessentially Darwinian." "A hundred years after [Darwin's] death," he concludes, "his approach seems more relevant to botanical studies than ever before" (Harper 1983:342). Let us mention here two further instances, one broad and theoretical, the other limited and empirical, of contemporary applications of Darwinian methodology, and even Darwinian ontology, to contemporary biological problems.

First, take Margulis's endosymbiotic theory of the origin of eukaryotes, first presented in her *Origin of Eukaryotic Cells* in 1970. Most evolutionary work is fairly specialized, dealing with the short-term variations of some particular living group or the detailed phylogeny of some extinct one. But here was a sweeping theory about the origins of all the populations that have ever inhabited this globe, combining Whittaker's five-kingdom classification with reliance on the phenomenon of symbiosis—a phenomenon whose existence no one could deny, but that Margulis holds has been unfairly neglected in evolutionary theory. Mere speculation, many believed. By now, however, the theory seems to be taken for granted. Thus Gould remarks, for example, of mitochondria: "They presumably originated, more than a billion years ago, as entire cells of primitive (prokaryotic) type that began living as symbionts within the ancestors of eukaryotic cells" (Gould 1987:14).

A host of evidence, Margulis insists, in the study of organelles—not only mitochondria, but also plastids—and in our increasing knowledge of Precambrian life supports this approach. Its acceptance has been hindered, she believes, because of the Darwinian emphasis on competition rather than cooperation in evolutionary theory (Margulis 1981). Yet her theory in fact appears as a striking exemplification of classic Darwinian reasoning, supported, of course, by immense advances in genetic, biochemical, and paleontological knowledge. In many circumstances, a hereditary symbiont has clear advantages over two free-living organisms that have to find one another in order to get what mutual benefit they can out of their complementary abilities: typically autotrophy and heterotrophy. On principle, there is nothing non-Darwinian about such a scenario. On the contrary: we have here, writ large, a classic case of the heritability of differential fitness. Some bacteria moved in to live permanently in the cytoplasm of other amoebalike protists and happily propagated themselves, as did their host organisms, in such a way as to give rise, in changing environments, to whole new populations of undreamed-of, ever-more-variegated plants, animals, or fungi.

Indeed, in terms of natural selection, the older view of the origin of higher plants was intuitively much less convincing than the endosymbiotic theory: to produce heterotrophy by losing the more direct economy of autotrophy seems less reasonable than to produce autotrophy in complex organisms through the chance—but happy!—assimilation of an already functioning alga by a heterotroph in an environment where such symbiosis may prove useful.

This is indeed a sweeping and speculative theory—and Margulis's additional belief in the symbiotic origin of the mitotic spindle may still seem dubious. But taken overall, her theory exemplifies magnificently the power of the general Darwinian framework to assimilate new evidence from an amazing variety of fields as well as new forays of imagination that probe far beyond the evidence available to Darwin himself without breaking apart the founding concepts of his vision.

For a second example of classic Darwinian thinking in contemporary biology, let us move from the very general to the very particular, to a case based not so much on the methodology of the *Origin* as on that of the *Descent of Man,* published in 1871. Michael J. Ryan, in *The Tungara Frog* (1985) explicitly applies Darwinian principles to the study of the mating calls of the males of a single species. He wants to know how female choice affects mating success, as compared with the increased risk of predation by frog-eating bats entailed by male calling. This is of course a case of sexual selection, which Darwin distinguished from natural selection, since it is not directly concerned with advantage for survival but with courtship behavior that may in fact prove disadvantageous for the immediate survival of the organism in question. Although this distinction was out of favor for a time, many evolutionists now favor its adoption (West-Eberhard 1979). We shall have a good deal to say about it later. At this juncture, however, we may take sexual selection as one form of Darwinian selection. All selection, all Darwinians recognize, has to do, indirectly or directly, with differential reproduction: survival mechanisms have no evolutionary effect unless they eventuate in the survivor's leaving more offspring than some less fortunately endowed conspecifics. Whether it is natural selection (for survival) or sexual selection (for obtaining or choosing a mate), it is selection of heritable variants with differential reproduction as its outcome that is in question. Of interest to us here is not this particular distinction, crucial though it will be to us in the following chapters. Nor are we concerned with the details of Ryan's results or methods. What we want to point out is the way in which he explicitly invokes not only Darwinian principles, but Darwin's very words, in order to explain the intent and procedures of his research, reserving an account of modern, technological complications for later. He writes: "Because explanations of some of the methods employed are rather tedious and are not necessary for an understanding of the results, only the general methods are outlined prior to the report of the results, and the more technical details are given in the appendix" (Ryan 1985:xiv).

Thus the appendix describes standard techniques for tagging animals in the field, statistical methods used, and, chiefly, technological devices employed in the course of the research to measure and analyze calls, to test female behavior in response to calls, to determine the energetic cost of reproduction, and so on. These are all up-to-date, late-twentieth-century procedures. The problem to which these procedures are to be applied, however, was set by Charles Darwin in 1859 and elaborated in 1871: the problem of how to assess female choice and how to measure its effects. Thus after quoting the *Origin* on the principle of natural selection, Ryan proceeds to quote Darwin's brief statement of sexual selection in the same work and then goes on to present the gist of the more developed treatment in *The Descent of Man* (Darwin 1871). He writes, quoting Darwin:

> There are two components to sexual selection: in the one it is between individuals of one sex, generally the male, in order to drive away or kill their rivals, the female remaining passive; whilst in the other, the struggle is likewise between the individuals of the same sex, in order to excite or charm those of the opposite sex, generally the females, which no longer remain passive but select more agreeable partners. (Darwin 1871:916)

So far Ryan quoting Darwin. One hundred and fourteen years later, Ryan continues: "Thus sexual selection results from competition and choice. Usually the competition is among males, and the choice is exercised by females" (Ryan 1985:3).

He then proceeds to expound further the problem of female choice as introduced by Darwin and as challenged or vindicated by modern biologists, including among the latter writers from Fisher ([1930], 1958) to Lande (1980, 1981) and Kirkpatrick (1982). What is probably rare in the history of science is to find contemporary problems and contemporary research set forth with recurrent reversions to the prose of a hundred years ago. Ryan's theory, for example, is backed up by another passage from *The Descent of Man:*

> In regard to structures acquired through ordinary or natural selection, there is in most cases, as long as the conditions of life remain the same, a limit to the amount of advantageous modification in relation to special purposes; but in regard to structures adapted to make one male victorious over another, either in fighting or in charming the

female, there is no definite limit to the amount of advantageous modification. (Darwin 1871:583; Ryan 1985:6)

New field techniques, improved statistical methods, and new technologies all contribute to the work; but the argument, the theory, and the problem are Darwin's, simon-pure after so many alleged revolutions: genetic, biochemical, molecular. Evolutionary biology, it seems, is still, at bottom, as Darwin shaped it.

Toward a New New Synthesis?

If Darwinian methodology works so well, then, why all our talk about its modification or expansion? Perhaps after all no new "new synthesis" is needed (Stebbins and Ayala 1981). The assimilation first of Mendelism, then of biochemistry, and most recently of molecular biology to the Darwinian perspective leaves the original theory undamaged. It can be applied either very broadly, as by Margulis, or in a different style, for example, by Kauffman (1992), or in very special contexts, as by Ryan (1985) and innumerable others (see, e.g., Bendall 1983; Grene 1981, 1983; Eldredge 1985). What more do we want of any theory?

The problem is, as we noted earlier, that, powerful as it is for guiding effective work in evolutionary biology, Darwinism lends itself recurrently also to distortions that weaken—though to some they seem to strengthen—its explanatory power. Since such exaggerated versions have had from the beginning (with "social Darwinism") especially regrettable consequences for our central area of interest in these pages, we need to look at these tendencies in a little more detail here.

It is from the ample but amply ambiguous character of Darwinism in its historical situation that the impoverished viewpoint we call ultra-Darwinism springs. Let us turn back to the ontological tensions we noticed in Darwin's theory, tensions that underlie the doubly "mechanistic" nature of his scientific vision. We listed three such polarities: determination and teleology, chance and necessity, and parts and wholes, or items and contexts. The second is an ultimate of Darwinian evolutionary thought, which it seems we must accept without further question as our starting point. The theory of natural selection is a two-step theory, as Sewall Wright, for example, described it succinctly and superbly in the 1967 Wistar Institute volume, and it is chance (that is, random variation in the evolution-

ist's sense of random—happening to happen, in no systematic connection with the needs of the organism), followed by necessitation
(that is, the resulting differential reproduction), that is involved.
The best resolution of the seeming paradox in this pair of concepts
can be found, perhaps, in Jacob's (1970) metaphor of evolution as a
"tinkerer." We just have to acknowledge that that is how life has
developed and continues to develop. It is the other pairs of contraries that make the trouble, especially in the ways in which they get
themselves entangled with one another. We have had some hint of
this in our introductory remarks about pseudoteleology and cryp-
toatomism in some types of hyper-Darwinian explanation—the kinds
of thinking exhibited, for example, in what Gould (1983) has called
the hardened synthesis. Let us look at this situation a little more
closely, or more broadly, in the light of our view of the historical
context of Darwin's thought.

Darwin is indeed the Newton of a blade of grass, not only in that
he brought living beings firmly within the purview of Western scientific thought, where previously they had held, if any, only a
peripheral position. He was himself, as we have seen, in some fundamental ways a Newtonian thinker. In particular, he provided for
the origin and maintenance of living things a thoroughly mechanistic causal explanation. Other things being equal, certain results
follow from interactions between organisms-slightly-different-from-
others and their environments (which include of course, other organisms). Even though such predictions are only probable, given
that numerous things might not be equal, and even though there is
always a stochastic element in the circumstances described or envisaged, given that the numbers involved are often large, this is still
straight, when-then, deterministic explanation. It is explanation *of*
adaptive arrangements, of means-end relations, and sometimes of
very elaborate and delicate machinery, but it is explanation *by* the
kind of straightforward one-thing-after-another causality beloved
of Cartesian-Newtonian natural philosophers. Such explanation,
however, has always been inclined to associate itself with atomism:
the belief that explanation should be, where possible, in terms of
least parts, which are the necessary components that "really" constitute what most of us see as wholes. If we get down to genes, we will
understand development; if we get down to neurons, we will understand thought. To deny the cosmic teleology of the theologians in
favor of naturalistic explanation seems, therefore, to mean that one
is looking not only for mechanistic cause and effect relations but

also for such relations between parts and parts, considering wholes as mere side effects. A hen is only an egg's way of making another egg.

Yet Darwin, like any working biologist—even the founder of population thinking—was fascinated not only by the slightly varying traits of particular organisms, pigeons, orchids, or earthworms but by the behavior of those organisms themselves consequent on their possession of slight differences from one another. And he was concerned especially with the behavior of organisms in their special "conditions of life." In other words, he was a deeply ecological as well as ethological thinker. Indeed, natural selection is inconceivable except in relation to environments: it is the changing scene, whether through selection in relation to a stable environment or in response (where the needed variation happens to be available) to environmental changes, that constitutes the history of life on earth. Eggs would be of no interest to a Darwinian evolutionist were there no hens, roosters, farmyards to forage in, or farmers (or farmers' wives) to raise and house and feed the poultry in question. The context matters, and matters essentially. To evolutionists who focus exclusively on the Newtonian aspect of Darwinism, however, the context consists of "vehicles" for the transmission from one body to the next of the parts that *are* the history, if history it be (Dawkins 1976, 1982). That is a grave distortion of the fuller range of Darwin's view.

Some recent critics, indeed, consider Darwin a more thoroughgoing Newtonian, taking his theory to be, from start to finish, an "equilibrium theory," working on a biological analogy of the principal of inertia (Depew and Weber 1988). This seems unfair. Some limited areas of Darwinian research, notably optimization theory, are indeed, as we have already noticed, equilibrium theories, in that they take the result of natural selection for granted and try to analyze the events that might have produced such results. But the background theory is more dynamic—a matter of wedges driving themselves in to disrupt a previous equilibrium. Moreover, if the calculations of population genetics take off from an equilibrium, what evolutionists do with them is to ask how that basic equilibrium might have been overturned, for clearly, where there has been evolution, it has been overturned . On the whole, Darwinism is rather a *dis*equilibrium than an equilibrium theory.

The problem, however—or one of the problems—is to sever the legitimate determinism of Darwinian explanation from an overzealous attachment to atomism or its biological equivalent. Of course analysis into parts can be, and often is, a major factor in mediating advance in the sciences, including the biological sciences—witness the recent achievements of work in molecular biology and its impact on systematics and other evolutionary disciplines. But when one is dealing with parts of organized systems, it is unwise to discard too hastily attention to the systems of which they are parts. Darwin did not make that mistake, but ultra-Darwinism does. There are wholes that need to be taken seriously in biology, from organisms to monophyletic taxa and from organisms to ecosystems to the whole biota (and the abiotic environment). If we miss them, we miss most of the subject matter of the biological sciences and, also, the enrichment of our concept of causality—but that is for later. All we want to do for now is to warn against the transformation of a causal concept, valid in its place, into a concept of causal connections "merely" between least parts.

If the association between mechanistic causality and reduction of wholes to parts produces misleading consequences, the assimilation of an atomistic thought style to teleology is even more bizarre. But it too is characteristic of ultra-Darwinism. Darwinian evolution is adaptive evolution: that's fine. There is nothing viciously (or anthropomorphically) teleological about the central role of adaptation, or adaptedness, in explanation by natural selection. True, a few contemporary Darwinians have tried to give the theory a kind of cosmic teleological slant by arguing that selection is *for* survival and in this is unlike the physical sciences, in which nothing is for anything (Ayala 1968). Such a claim, which issues inevitably in the old pseudoproblem of evolutionary tautology (what survives survives!), is easily defeated (Grene 1974).

The mode of teleological description characteristic of a careful Darwinism (or its teleonomy, as Mayr [1974] prefers to call it; cf. Mayr 1988) presents no such difficulties. But, again, the trouble arises from combining the telic phenomenon of adaptedness with an atomizing thought style. Selection selects small variations: true. It is slight differences from which evolution starts. And selection selects only what is useful (in the ultimate context of differential reproduction). Therefore, if there is selection, some little traits, some minute differences, must be "worth" propagating. It does *not* follow, however, that *all* minute traits are in themselves adaptive. Yet, given the adaptation-machine model, it is easy to slip into that invalid inference and to take every minutest part of every organism as in itself and necessarily good *for* something. In a discussion of the power of selection to make use of a variety of genetic factors, the great cotton geneticist Harland once remarked that there were thirteen alternative genetic determinants for a small orange spot on the petal of the cotton flower. Someone asked him what the spot was good for. No one knows that, he replied, but of course it must be good for something, or it wouldn't be there.

Darwin discussed at length in *The Descent of Man* characters that in his view were merely "chemical," irrelevant to selection (in our terms, "neutral"), but ultra-Darwinians refuse to accept such pluralism in their ontology. Every little something must be good for something, or things wouldn't be made of nice tidy bits, and selection wouldn't be wholly and always in control of all of them, as of course we know it is. So we have a pervasive teleology linked with atomism, and at the same time a stock, one-level determinism linked with the same compulsion to atomize. It is this combination that produces the impoverishment of Darwin's theory in its ultra-Darwinian form.

Behind these tendencies, in turn, lies the lingering force of what was once "the new mechanical philosophy." To elude its spell and to maintain and/or expand the richer biological insights inherent in Darwin's evolutionary vision, we need to articulate an enriched ontology for the biological sciences, an ontology that not only "restores the organism to biology" (Gould 1980a), but also restores the environment to its pivotal place in evolutionary theory. Furthermore, it should firmly put aside an exclusive emphasis on least parts and recognize the existence and the causal roles in evolutionary history of entities not only beyond the gene, but beyond the organism as well.

CHAPTER THREE

Entities, Systems, and Processes in the Organic Realm

Social behavior, like behavior generally, is an aspect of the phenotype—taken here as specifiable attributes of organisms. Such attributes are generally distributed uniformly enough so that they can be recognized, and thus compared, among a group of organisms. Phenotypic properties include physiological, anatomical, and behavioral features. Indeed, the first two are the more traditionally considered categories of phenotypic properties in evolutionary biology, simply because there has historically been less difficulty in establishing the genetic basis of nonbehavioral, anatomical phenotypes; and heritability is a sine qua non in evolution. Evolution has come to mean the history of genetically based information within lineages. And most genetic information concerns the differentiation, growth, and maintenance of physical structures, physiologies, and attendant behaviors of organismic phenotypes.

Yet it would be a mistake to insist upon a genetic basis— or even a strong, if incomplete, genetic component—of every item of comparable, varying social behavior before such are seen to have had evolutionary histories that can be approached in a biological context, for that context is supplied by the moment by moment functions and processes that characterize all organismic life: the functions and processes (behaviors in the most general sense) that serve to maintain the soma of multicellular organisms and that operate, as well, in the reproductive transmission of copies of the underlying genetic information to form the next generation. As long as both the basic, dual processes of (1) organismic somatic differentiation growth and maintenance (which, as a set, amount to matter-energy conversion processes)—*and* (2) reproduction with underlying replicative fidelity of the genome are in place, the context is set for an evolutionary consideration of behaviors, social or not, genetically based or not. Regardless of means of transmission, social behaviors arise as an outgrowth of—and, most importantly, as a bridge between—the economic and reproductive activities of organisms.

Gould and Vrba (1982; see also Bock and Von Wahlert 1965) have recently clarified the taxonomy of features, functions, and roles played by organismic phenotypes in an evolutionary context. In so doing, they have helped to tease apart starkly contrasting aspects of adaptation. Their primary point is that such functions observed in operation in the modern biota may not have been shaped by natural selection to serve the particular role observed; the feature may simply have been co-opted, that is, found to serve a useful purpose that is unrelated to the original use for which it was evolutionarily shaped. This evolutionary history of such a feature *for the new use* is best understood as an incidental side effect—a usage that flows from Williams (1966). Thus their taxonomy: "aptations" are phenotypic attributes that enhance fitness (statistical probability of successful reproduction); "adaptations" are those features shaped directly by natural selection to fulfill the role or purpose they are observed to serve; "exaptations" are those co-opted because of their chance utility for a role not directly shaped by natural selection. The aye-aye, *Daubentonia madagascariensis,* uses its anatomical specializations, most notably, ever-growing incisors and an elongated fourth digit on the manus, respectively to penetrate coconut husks and to scoop out the internal meat—an evident exaptation, since the aye-aye anatomical and behavioral feeding package originally served exclusively as a means of disgorging insects from tree bark in a precoconut Madagascar environment (Tattersall 1982; pers. comm. 1989).

We would like to go a bit further. To pronounce obviously useful, functional phenotypic properties as aptations is an improvement over the assumption that such properties necessarily represent true adaptations. Yet there is a residual difficulty even with the term *aptation*. For there has been an elision in our approach to current functions as well as to their history in evolutionary biological discourse. The tension in evolutionary theory between the two great classes of organismic biological process (i.e., between economics and reproduction or genealogy) has contributed to the confusion between history and current utility (including, but not restricted to, the confusion discussed by Gould and Vrba 1982) and to an even more critical misreading of the organization of biological systems that further obscures the evolutionary context of social behavior.

Past and present conceptions of fitness epitomize the problem. To Darwin and his contemporaries—as well as to laypersons today—fitness means something like robustness and vigorous good health. To a population geneticists, and hence to evolutionary theorists ever since the 1930s, fitness means specifically the relative probability that an organism will reproduce successfully—relative, that is, to conspecifics in a local population often, but not necessarily, of the same sex. Geneticists are well aware that many factors having nothing to do with general good health (or, specifically, of the relative goodness of a particular phenotypic trait, a small subset of somatic features) militate for and against any one organism's reproductive track record. Thus fitness is rendered as $1-s$, where s is a "selection coefficient," which may be quite small. Even if only 3 organisms out of 100,000 per generation have depressed reproductive success because they carry unfavorable versions of some trait, that selective pressure can amount to considerable genetic change as evolutionary time marches on.

There is nothing inherently wrong or vague about (re)defining fitness as relative probability of reproductive success. Yet what is lost in so doing is the old, clear distinction between economic,[1] somatic, organismic vigor and reproductive activity. Darwin (1859) was well aware that no straight, absolutely certain vector connected health and vigor with reproductive success. The demonstration by mathematical population geneticists that even slight selection pressures can effect significant genetic change has served to strengthen Darwin's claim of a direct causal connection between economic and reproductive success. But in redefining fitness as probability of reproductive success, we in effect erase that thin line, bringing vigor—fitness in the old and still colloquial sense—cheek by jowl with reproductive success. We *mean* "only to the slight extent that the two are correlated." But the way is paved, as we have already outlined in chapter 2, for seeing the purpose of any phenotypic trait to enhance reproductive success. The economic role, of say, sharp teeth may be the more efficient capture of prey; but those sharp teeth were honed, in an evolutionary sense, because they caused differential reproductive success, and thus their underlying genetic information was differentially preserved. Current economic role and evolutionary history emerge as one and the same thing—the current state of affairs in much of gene-centered modern evolutionary theory, variously regarded as something of an intellectual triumph (e.g., Dawkins 1986) or an impenetrable conceptual jumble.

Physiologists have long been aware that, of the spectrum of organismic functions traditionally falling under their purview, reproduction is unique in not being essential to the survival of the organism itself.[2] Reproduction, instead, costs energy. Thus reproduction, inessential to an organism's life, is generally the first to go when times become rough—when, in general, there is insufficient energy capital to expend on the luxury of reproduction. Even in humans, whose societies reflect such a tight and labyrinthine reintegration of sexual and economic activities, reproductive dysfunction commonly results from general somatic stress—as in the well-documented tendency for disruption of menstrual cycles in women athletes undergoing intensive, rigorous training.

Thus there is an inequality between the two sets of organismic activity—between reproduction on the one hand, and respiration, excretion, digestion, and all the other intra- and extracellular physiological functions. Reproduction requires a source of energy, as well as a somatic "house"—a casing for the genes that serves as their protection, structural support, and source of energy. To exist and to function, the somatic, economic side of the organism does not need reproduction. Mitotic cell division with attendant genome replication *is*, of course, the very essence of maintenance, growth, and differentiation of the soma, a vestige of phylogenetic days when the distinction between economics and reproduction was far less clear-cut than it is in multicellular organisms of the past billion years or so of earth history. The distinction between these two classes of biological activity is valid only when by "reproduction" we mean the "making of more organisms of like kind." We shall document the increasing split between reproductive and economic processes within organisms and within biological systems in general later in this chapter. It is this dichotomy that sets the stage for an evolutionary, and distinctly biological, understanding of the nature of all social behavior. But before considering the distribution of these classes of activity in greater detail, it would be beneficial to underscore a bit more vividly the obvious evolutionary import of such a distinction. For it is Darwin's grasp of this distinction that underlies the original formulation of natural selection—and his insistence that natural selection and sexual selection are distinct processes.

Natural and Sexual Selection

Darwin (1859, 1871) realized that many factors influence reproductive success. The diversity of life (meaning phenotypic, rather than taxic, diversity—yet another distinction all too frequently glossed over in evolutionary biology) formed the central issue to be explained in Darwin's original formulation. It remains primary today. Darwin saw nature acting much as the selective animal husbandryman sees it: knowing that domestic and natural populations vary with respect to phenotypic traits, and that organisms tend to resemble their parents, Darwin simply imagined that relative success in putting those variable traits to use—or, rather, the relative economic success resulting from the variation in phenotypic traits— would show up, ceteris paribus, as reproductive success. How well an organism does in its economic life in the prosaics of staying alive, perhaps even flourishing, would on the whole enhance that organism's chances of reproducing successfully.

It is often remarked (as we have already seen in chapter 2) that selection is a two-step process: there is variation, the ultimate source of which is now known to be mutation; and then there is selection proper, where "the environment" determines (albeit stochastically) which of the variants will increase relative to the others. Yet in natural selection, at least as originally framed, this second step itself has two critically different aspects: there is relative economic success, which may lead to a differential bias in reproductive success; but there is also reproductive success that is *not* dependent upon (relative) economic success. That Darwin kept economic success distinct from (and therefore one of the causes of) reproductive success is abundantly clear from his original definition of sexual selection.

Darwin (1871:256) saw sexual selection as arising from "the advantage which certain individuals have over other individuals of the same sex and species, in exclusive relation to reproduction." One possible complicating factor in the equation "differential economic success = differential reproductive success" is simply that some organisms (of, as Darwin said, the same sex and species, and living in the same local population) may simply be better at reproducing than others, owing purely to a superiority in one or more aspects of reproductively centered phenotypic features. Nor are the moral overtones of such formulations hard to spot: rather than receive the (just) rewards of a lifetime of economic success (and perhaps hard work—certainly physical superiority), some organisms within a population "cheat" and, having in any case sufficient resources, manage to reproduce in spite of their manifest lower success in the economic sphere.

As in so many formulations in evolutionary biology, sexual selection (group selection is another) is often seen to exist if and only if it *opposes* natural selection (as in the sneaky-fucker scenario). Epistemologically, the point has some force: it is easiest to establish an alternate form of selection if the example at hand shows conflicts. Just such an antagonistic relation was said to obtain in the only example of group selection accepted by Williams in 1966—in the lethal t-allele in mice, which spread because of some putative good to the group, despite the lethal effect of the locus in homozygous state in individual organisms. Yet there is no reason why, in principle, the directions of different modes of selection have to oppose one another (Wimsatt 1980).

More interesting, and as we have already seen, sexual selection has recently begun to come back into favor as a truly distinct mode of selection—distinct, that is, from natural selection. And it is important to ask why most of us were taught that sexual and natural selection represent distinctions with no real difference, and why many of us still prefer to go along with the Darwinian revisionists who have denied any real difference—by, for example, maintaining that natural selection embraces both sexual selection and some as-yet-unnamed subcategory that reflects an impingement of differential economic success on reproductive success. Yet the latter category is what Darwin originally called natural selection.

The blurring of Darwin's original distinction in modes of selection parallels the elision already seen in the use of the term *fitness* over the past thirty years. There simply seems to have been no evident use to which the distinction could be put in theory—and the blurring of economic and reproductive success smoothed over the uncomfortable lack of precise determinism, that inherently sloppy relationship, between economic and reproductive success. The realization that sexual selection is a real phenomenon, as some sociobiologists are discovering, stems partly from a recognition that reproductive success can be high even when economic success appears to be low. But even here, the discourse remains somewhat at cross-purposes, for the ironic reason that, to many sociobiologists, sexual selection is synonymous with mate selection—meaning mate "choice." We will follow Darwin (1871) in a more extended, general conceptualization of sexual selection: in our view, sexual selection is responsible for all aspects of the development and maintenance of phenotypic features pertaining directly and primarily to reproduction—that is, reproductive adaptations. Such most especially include all elements of the Specific Mate Recognition System (Paterson 1985), as we discuss in greater detail in chapter 4.

The core of the distinction between economic and sexual selection resides in Darwin's phrase "in exclusive relation to reproduction." It emerges, further, in Weismann's distinction between the germ line and the soma. For in all organisms—with the crucial exception of some social and colonial organisms—there is anatomical, physiological, and behavioral machinery for economic functions and distinct machinery for reproductive functions. The distinctions are clearest, of course, in multicellular organisms. There is, in short, a distinction, sharpest in the most complex of organisms, between a large class of economic (somatic) properties (phenotypic attributes) and reproductive (genealogic) properties. To the extent that it is appropriate in any instance to refer to the historical development of such properties as adaptations, the distinction that Darwin drew amounts to the recognition of two classes of adptations: economic (somatic) adaptations and reproductive (genealogic) adaptations.

It is the uses to which these two different classes of adaptation are put by organisms that causes organisms to be integrated into two utterly separate systems—one economic, the other genealogic, both hierarchically arranged. It is the existence of the genealogical hierarchy (consisting of demes, which are parts of species, which are parts of monophyletic taxa) distinct from the economic (ecologic) hierarchy of populations (avatars; Damuth 1985), which in turn are parts of ecosystems—it is this division that sets the stage for the reintegration of economic and reproductive organismic adaptations in social behavior and social systems. Before turning to a detailed consideration of the distribution of organismic adaptations in two separate hierarchical systems, we shall briefly consider the phylogenetic increase in the split between genealogical and economic somatic processes as organisms have become more complex.

Weismann's Doctrine and the Burgeoning Split

Natural selection, in its pure, original, Darwinian conceptualization, represents a vector of causality: what happens to conspecifics within local populations as they conduct their economic lives has a statistical effect on relative reproductive success. Economics determines to some degree what genetic information is passed on to the next generation. Although the bulk of that information concerns economic matters—that is, the hereditary basis for phenotypic attributes concerned with economic functions—the actual process of passing that information along is, of course, strictly genealogical. At the among-organism, within-population level, economics exerts control over the genealogical processes of transmission of genetic information.

Not so *within* organisms, at least insofar as multicellular organisms are concerned. Weismann's dictum—that the germ line determines the soma, but that the converse, where the soma exerts some causal impact on the germ line, does not obtain—effectively runs in exactly the opposite direction to the "selection vector." Of course, it is the soma that houses the reproductive organs and cells, that supplies the energy for replication and transmission of the gametes, and so forth. Hormonal and neuronal pathways further integrate reproductive morphologies, behaviors, and physiologies with those of the soma. Yet in an evolutionary context (and despite recent attempts at falsification; see Steele 1979), Weismann was correct: the germ line produces the soma; what happens to the soma has no effect on the hereditary factors—the genetic information—of the germ line. And it is the transmission of that information in both conserved and modified form that constitutes evolution. Insofar as evolution is concerned, at the within-organism level, it is the genetic processes of the germ line—the genealogical side of the ledger—that take precedence over the somatic, or economic, side.

Molecular biology has only reinforced the late-nineteenth-century generalization of August Weismann. The "Central Dogma" asserts the primacy of nucleic DNA, which sends its instructions out to various areas of the extranuclear cytoplasm to serve as templates for protein assembly. Retroviruses and reverse transcriptase notwithstanding, it is the norm in all eukaryotic and prokaryotic cells that the structure of those economic molecules—proteins—is not translated and transported back to effect modification of the structure of nucleic DNA. It is a biochemical one-way street, whether

DNA is replicated in the germ line to produce haploid gametes that will then fuse with another, forming a zygote that then proceeds to grow and differentiate into a descendant multicellular organism; or a single-celled eukaryotic or prokaryotic organism simply divides (whether or not following an exchange of DNA with another organism) to produce an offspring; or a somatic cell of a metazoan divides for growth or simply somatic maintenance; or RNA transcribes the DNA sequences in conjunction with intracellular (but extranuclear) protein synthesis. The information of the nuclear DNA determines the structure and content of its copies, be those copies more DNA, or the various RNAs and protein end products.

This asymmetrical vector of causality has strongly influenced much of the theoretical study of the origin of life ("biogenesis"). Because proteins are assembled through the chemical templates of RNA, and DNA commonly underlies the informational system culminating in protein manufacture, it is natural to hypothesize (or at least to imagine) that 'twas ever thus—that DNA and RNA preceded proteins in life's history.

Reinforcing such an expectation of the primordial primacy of the molecular basis of heredity is the observation that the economic activities of organisms—the acquisition of energy and its use in the construction of large, complex structural molecules—are by no means unique to "life." Matter-energy transfer processes, including the construction of "dissipative structures," are wholly characteristic of the material universe in general—no matter how much the specific form of such processes might vary from system to system. But exact replication of the structure of a molecule that serves variously as the template for an entire new organism, or simply for the construction of specialized molecules needed to maintain the physiology of the parent cell, is, of course, unique to biological systems. Small wonder, then, that DNA and RNA are often seen as the sine qua non of life—and therefore as *the* phenomena to be confronted when the origin of life is considered.

Yet the fact that matter-energy transfer processes are more generally distributed in the material universe—that is, are by no means confined to the subset "life"—is no reason not to consider the origin of proteins simultaneously with the origin of the macromolecules of heredity. Experimentalists have realized for years that amino acids form spontaneously, given the proper initial conditions; and the whole point of the famous Miller experiments was that short-chained proteins are readily assembled from such amino acids, given but a few additional required conditions (and preconditions). As Shapiro (1986) and others among the theorists have also argued, there is really no reason to see the Central Dogma as primordially valid: why couldn't RNA, say, have been "invented" as a chemical matching of bases with the amino acids of spontaneously assembled protein skeins?

Our goal here is to stress neither side of what seems to be shaping up as a classic chicken-and-egg dilemma in biogenesis. The point is that all but a few specialized cells (whether free-living, as entire organisms, or specialized parts of multicellular organisms) make more cells of like kind (at least as adults; the two-hundred-odd cell types of a multicellular organism such as *Homo sapiens* all differentiate from a single cell—the zygote) and lead an economic existence, that is, require a source of energy to maintain themselves and perform whatever metabolic functions are peculiar to them. All organisms, from the first replicating cells onwards, have always been dual entities: they are economic machines, and they are reproducers, making more entities of like kind, with replicative fidelity supplied by the underlying genome. Neither set of processes is more important than the other in specifying what life actually is. The two sets of processes are equally fundamental, equally functionally necessary, and thus inseparable from any consideration of the nature— or origin—of life.

Yet as inseparable as they may seem, these sets of organismic processes have actually gone a long way towards separation as the phylogenetic history of life over the past 3.5 billion years has proceeded. We have already stressed, at the outset of this chapter, how the economic and reproductive activities of organisms are decoupled enough so that natural selection arises as a virtual necessity. It is only because organisms lead largely separate economic and reproductive lives that relative success in the former leads (under a few necessary conditions) to relative success in the latter. (For organisms in general, the reverse does not obtain.) But in addition to the evident separation of the two grand classes of biological process, which leads directly to natural selection, there has been a trend to an actual *physical* segregation of these two activities within organisms.

For the past twenty years, Whittaker's (1969) scheme of five kingdoms has dominated conceptions of the major taxa of living organisms and their interrelationships. Yet it is noteworthy that his formulation repeatedly sees life and its various subdivisions divided into opposing moieties—the "haves" and "have-nots." As cladistics has made abundantly clear in recent years, it is relatively easy to recognize and define monophyletic taxa on the basis of shared possession of organismic phenotypic features that represent evolutionary novelties. Indeed, clearly delineated higher taxa are nothing more (nor less) than lineages made prominent by the appearance and retention of easily observed and generally uniquely held organismic characteristics that serve as "markers" (M. J. Novacek, pers. comm. 1985) for phylogenetic systematists. More difficult are those taxa left over in the analysis, that is, those that lack the specializations that single out the advanced subset. Traditional practice sees nothing wrong with defining such have-nots as a coordinate set, generally ranking equally (i.e., in the Linnaean hierarchy) with the haves. The problem with such coordinate ranking is that the have-nots are not defined by any comparable set of morphological features not present in the advanced subset (i.e., the haves); thus there is no way to be sure if, or in what particular specific way, the members of the have-nots are themselves phylogenetically interrelated.

Two "superkingdoms"—the Prokaryota and the Eukaryota—exist in Whittaker's five-kingdom scheme. Prokaryotes are bacteria, while eukaryotes are everything else. Bacteria are universally considered the most primitive kind of true organism still living; eukaryotes are everything alive *other* than bacteria.[3] As the names imply, they are distinguished by details of cellular anatomy—the "pro"karyotes represent the incomplete configuration of true or wholly formed "eu"karyotes.

Margulis (1974) lists features of cellular anatomy that serve to characterize eukaryotic cells; we draw on that list, with the corresponding "defining" features for Prokaryota, in table 3.1. Note that for most of the defining characters of Eukaryota, the corresponding defining features for Prokaryota is *absence* of the eukaryotic characteristic. As generally proves to be the case with large taxa defined on the basis of retention of phylogenetically primitive features, intensified research often demonstrates previously unsuspected heterogeneity within such plesiomorphic taxa, and the bacteria are no exception. The work of Woese (e.g., 1981), in particular, has demonstrated a rich diversity, especially in metabolic pathways, within bacteria in general. And there is little doubt that among living bacteria, some are more closely related to eukaryotes than are others. Dividing organisms into sets of haves and have-nots obscures

TABLE 3.1.
Defining Properties of Prokaryotes and Eukaryotes

Prokaryotes	Eukaryotes
Small cells	Large cells
Nucleoid not membrane-bound	Nucleus membrane-bound
No centrioles, mitotic spindle or microtubules	Centrioles generally present, mitotic spindles, microtubules present
Sexual system absent	Sexual system generally present
No mitochondria	Mitochondria
No intracellular movement	Intracellular movement

After Margulis (1974), table 3.

lines of descent and conveys the false impression of phylogenetic cohesion to the have-nots.

Yet even though we cannot take Bacteria seriously as a natural, phylogenetically unified group, they do constitute a "level of organization" (see, e.g., Schaeffer 1965), representing in general the condition from which the derived, phylogenetically undoubtedly unified Eukaryota arose. And if we focus on those features of cellular organization that unite eukaryotes as a monophyletic taxon, it is immediately apparent that the overall increase in complexity of cellular organization reflects an increase in division of labor. And that labor covers both categories of biological process.

Margulis and Sagan (1986) have recently underscored the lability of the prokaryotic genome. Far from not indulging in "sex," bacteria are almost wildly promiscuous in the exchange of snippets of DNA. Nor is such exchange limited to conspecifics; Margulis and Sagan paint instead a picture of an almost conspiratorial world of the microcosm, where useful innovations are widely disseminated among lineages that have been phylogenetically distinct for billions of years. For example, they report (Margulis and Sagan 1986:91) that the initial mutation conferring resistance to penicillin in staphylococci probably arose in a lineage of soil bacteria and simply spread to staphylococci, which could put it to obvious good use.

Things are apparently not so labile in the eukaryotic genome—partly, at least, because loss of flexibility often seems the price of increasing organization and complexity. Just as there is a correlation between anatomical complexity and loss of ability to regenerate differentiated organs in metazoans (even within the Vertebrata), the genome of eukaryotes appears to be more restricted in its ability to recombine with genomes of other (and especially only remotely related) organisms. In particular, the physical segregation of the genome into a discrete nucleus, bound by a double-walled membrane, signals the double-sided task of what in eukaryotes has become truly "nucleic" acid. Organization of DNA into discrete chromosomes both regularizes the process of reproduction (in its most general sense, i.e., true organismic reproduction of single-celled creatures, or mitotic somatic cell division, or meiotic germ line cell division in multicellular organisms) and further limits whatever biochemical bounds to DNA reorganization are already present in prokaryotes.

But the invention of a discrete nucleus also serves to segregate more precisely the information-bearing molecules from the sites of physiological economic activity that now (i.e., within eukaryotes) take place in a differentiated cytoplasm. The machinery whereby DNA sequences are translated and transported by RNA reflects a literal, physical distancing between genetic information and its use in carrying out the economic business of a cell. If the DNA is forever destined to remain with the dual role of supplier of information for protein assembly *and* source of copies for the production of descendant organisms of like kind, nonetheless the two activities become much more sharply defined and differentiated in eukaryotic than in prokaryotic cells.

Moreover, with the appearance of eukaryotic cells, there is an equally obvious cytoplasmic differentiation of intracellular organelles devoted to a variety of specific economic tasks. Not only are reproduction and economics more formally separated, but economic (physiological) roles are likewise allocated to specific sites. Presumably such innovations represent efficiencies in an engineering sense, enhanced use of energy sources that set the stage for still further modifications in the use of economic resources. The once-provocative (but now largely accepted) hypothesis advanced by Margulis (e.g., 1974) sees many of the intracellular cytoplasmic organelles as the phylogenetically incorporated (literally) results of ancient symbioses among various prokaryotes. Her hypothesis envisions an assembly of disparate parts rather than an evolutionary differentiation as the source of economic complexification as eukaryotes arose from prokaryotes. In any case, what matters most in the present discussion is that such subdivision of labor—of reproduction from economics, and between various subsets of (economic) physiological processes—did in fact occur as eukaryotes arose from prokaryotes.

The eukaryotes themselves present a second striking example of the haves and the have-nots. Whittaker (1969) sees four eukaryotic kingdoms: the unicellular Protista and the separate, multicellular Fungi, Plantae, and Animalia. Whereas the list of features that characterize eukaryotic cells appears to nearly everyone's satisfaction to represent perfectly good synapomorphies (shared evolutionary novelties, thus prima facie evidence of eukaryotic monophyly), no one supposes that multicellularity constitutes a comparable synapomorphy uniting plants, fungi, and animals into a monophyletic taxon. Rather, they are all reckoned to have arisen separately from protist ancestors. And, indeed, from within the vast array of different protistan taxa, there are some with manifestly close ties, variously with plants, animals, and (less clearly?) with fungi. Protists are even more obviously a phylogenetically heterogeneous "taxon" than is Prokaryota, united only by their retention of the primitive feature of unicellularity.[4]

Protistans, as the primordial eukaryotes, represent the level of economic and reproductive functional differentiation that we have already seen in contrast with the prokaryotic condition. Multicellularity, on the other hand, represents still further differentiation among economic functions per se, and between economic and reproductive activities generally. There is, of course, an economic differentiation along phylogenetic lines: animals are heterotrophs, entirely dependent, at base, on the photosynthetic activities of the autotrophic plants (as well as on various autotrophic protists and bacteria); the saprophytic fungi are yet another matter. But more to the point, there is further economic differentiation within the soma of single multicellular organisms. And there is also, as Weismann was the first to make clear, that all-important distinction between the economic soma and the reproductive germ line.

We will confine this brief characterization of multicellular organisms to animals. For one thing, germ layers differentiate from somatic tissue in plants, blurring Weismann's distinction. And it is animals, of course, that form social systems. Yet it is clear that some plants (if not fungi) exhibit a considerable degree of economic differentiation and can even be considered formers of colonies (if not full-blown social systems).

Within Animalia there is a rather complete spectrum of multicellular organisms. The Parazoa (sponges), which possess but a few cell types and no truly developed tissues, stand as the sister taxon to all true animals. Cnidarians (corals, jellyfish, and their relatives), possess a double layer of tissues, integrated with a simple nerve network; they lack any true organs. Flatworms and other acoelomates[5] show the development of true organ systems—including gonads that produce meiotically haploid gametes. It is at this, still relatively primitive, level within multicellular (i.e., true) animals that the separation of germ line from somatic organs systems becomes marked. And thus Weismann's whole-organism equivalent of the molecular Central Dogma becomes a matter of great evolutionary significance.

The scale of complexity increases to the point where some organisms have over two hundred cell types—not just *Homo sapiens*, but mammals in general. Arthropods, mollusks, and echinoderms also exhibit a high degree of somatic differentiation. Crustacea, for example, typically develop (especially in the "higher" orders) a highly differentiated series of head and trunk appendages, modified to serve a variety of functions, including different feeding tasks, locomotion (both walking and swimming), respiration, defense, reproduction, and so forth. Yet all organisms, no matter how complex, continue to lead economic and reproductive lives—with some interesting variations and even exceptions that, if anything, dramatize the distinction between general economic and reproductive organismic morphologies and behaviors.

Norms and Departures: Reproduction and Economics in Complex Metazoa

With the growth in complexity attendant upon multicellularity, the soma/germ-line distinction amounts to a very real, if unbalanced, apportionment of energy resources, time, and effort. The vast bulk of most organisms' lives is devoted to acquiring energy and using it to grow, differentiate, and maintain the soma. Some of that growth, maintenance, and energy utilization is occasionally expended on reproduction. These statements, of course, are intended to be inclusive of organisms in general; they are reasonably accurate for most Metazoa, as well, though behaviorally complex higher vertebrates may seem to belie the generality for reasons that will become clearer in later chapters. It should be borne in mind, however, that much of the recent work in behavioral ecology takes sociobiology as its theoretical matrix. And, as we have discussed at the outset, sociobiology is a logical outgrowth of Darwinian biology, particularly the ultra-Darwinian version of neo-Darwinism, which tends to see competition within local populations of conspecifics for the differential spread of genes as the role and evolutionary "purpose" of all adaptations. Thus, in sociobiology, the organizing purpose behind social behavior is the struggle to spread the genes underlying the behavior—thus biasing observations and reports of animal behavior towards the reproductive side of the ledger (see Oster and Wilson 1978 for a refreshing exception to this statement). Even most social animals devote relatively little time *directly* to reproductive matters—though extended interactions between economic and reproductive behavior beyond those envisaged in the pure version of natural selection really constitute the biological basis of social behavior.

This being said, there are many exceptions to the bald statement "All organisms spend a good deal of time pursuing economic affairs, and some smaller amount of time reproducing." Nor, for the moment, need we consider nonreproducing social castes, whose role(s) in reproduction are nonetheless established adequately in sociobiology by construing their economic behavior as an adjunct of the reproductive activities of another caste (through kin selection; see chapter 6 for further discussion). Rather, at this juncture we need to briefly consider whether there are any major exceptions to the dual economic and reproductive lives of organisms in general by considering nonsocial (and noncolonial) organisms.

Organisms with true sexual reproduction, where (as a rule, haploid) gametes from different sexes fuse to form a zygote, constitute the only class of organisms among which there are some significant departures from the norm. And the departures are one-sided: no organism can avoid being a matter-energy machine to some extent (though some organisms never feed; e.g., male mosquitos, who feed as larvae, but who as imagos use stored energy resources to maintain their bodies, fly, and mate); and, somewhat ironically, though reproduction is the one function an organism can successfully dispense with (even might need to avoid, simply to conserve energy to survive), reproduction cannot be eliminated in evolution (for reasons already discussed) except in some social colonial organisms, where the rules have been changed. Asexual castes and members of colonies live in situations in which the division of labor is even greater than within the bodies of single, complex metazoa; kin selection is particularly effective in clonal situations, where asexual and sexual organisms are genetically identical, as in the zooids of bryozoan colonies.

Yet the one-sided departures from the norm are toward, rather than away from, reproduction. Some organisms seem little more than sex machines, so reduced is their role in economic life. A particularly graphic example occurs in the sexual dimorphism of the paper nautilus, genus *Argonauta*. According to Morton (1958), males of some species are reduced morphologically to being little more than a hectocotylus—in normal male squid morphology, a modified arm ("tentacle") used for transferring sperm packages to the female mantle cavity. In *Argonauta,* the male, far from living life as a typical independently predaceous squid, is instead many times smaller than the female and is little more than a reproductive organ.

In other instances, both sexes seem bent on only reproduction throughout their entire lives: mayflies, with their ephemeral adult lives of a single day, seem an obvious case in point. But even mayflies lead economic lives, however brief (especially to our perspective; mayflies undoubtedly experience it as a lifetime). And though it cannot, seemingly, remove death, selection quite obviously does set average life spans, as well as timing of reproduction within that life span. So, although mayflies appear at first to be an example where both sexes spend most of their available adult lifetimes pursuing reproduction, they are perhaps more accurately seen as a species where adult economic lifetimes are reduced to a minimum, vis à vis the norm for other insects.

Examples such as the male *Argonauta* do tend to bolster the supposition that, somehow, reproduction is the ultimate purpose of active life. We have no real counterexamples—where reproduction is put at some irreducible minimum, and long and purely economic lives are led.[6] Yet it must always be borne in mind that the economic life of an organism (usually males of a species), though curtailed, can never be wholly eliminated. However unbalanced the ratio of activity has become in some taxa, all free-living (noncolonial), including at least some forms of social, organisms are reproducers and economic machines. And it is the commingling of generally separate economic and reproductive functions of organisms that has led, in various taxa at various times, to the emergence of social behaviors and social systems.

Using Those Adaptations: An Introduction to Organisms in Groups

Evolutionary theory has always focused on the origin and further modification of phenotypic adaptations. Though an explicit realization that there are two general classes of phenotypic features, hence of adaptations, has been nearly lost in modern versions of Darwinism, it was definitely there in his original formulation. And, as Hull (1980) has clearly pointed out, it is the differential economic success that underlies differential reproductive success to yield natural selection—and the origin and further modification of economic adaptations. In contrast, sexual selection accounts for the origin and further modification of phenotypic properties exclusively related to reproduction—properties subsumed in Paterson's (1985) Specific Mate Recognition Systems.[7]

We have briefly described how the simplest of organisms show little physical differentiation of morphologies and loci of processes between the two classes of biological activity; how eukaryotes demonstrate both a greater separation of the two as well as a definite tendency towards subdivision of purely economic labor within the cell's cytoplasm; and how multicellular eukaryotes—with particular emphasis on Metazoa—carry both tendencies still further. With Metazoa, we find Weismann's doctrine materialized in a sharp distinction between two general classes of organ system; as the Central Dogma tells us, the asymmetrical vector—in which genealogical processes are paramount over economic ones, within organisms, in an evolutionary context—holds for virtually all organisms (with only some colonial and social organisms excepted). And, finally, we have seen how (natural, or economic) selection reverses the Weismannian vector of causality: it is relative economic success among organisms within populations that determines, in some fractional measure, what genetic information is to be passed on to the succeeding generation.

Yet there is more to the picture of biological entities, systems, and processes than organisms alone, for there are direct consequences of the possession of these two sets of adaptations. Two distinctly different—and increasingly divorced—sets of associations of organisms arise as an automatic consequence of these two distinct sets of behaviors, physiologies, and morphologies housed within nearly every single organism. Social systems are aggregates of organisms. Understanding the nature of the groups that organisms in general form reveals the actual biological nature of the special systems—specifically, social systems.

The economic behavior of organisms in groups forms the traditional subject of ecology. Systematists, on the other hand, examine the patterns of genealogical interrelationships among species and monophyletic taxa; as we shall see in the next chapter, these latter groups form automatically as a consequence of reproductive behavior, most sharply distinct in sexually reproducing organisms. Species and monophyletic taxa are natural outgrowths of simple reproductive behavior; local and regional ecosystems, in contrast, necessarily arise as a simple consequence of the economic activities of organisms. In ecosystems, matter-energy transfer (and related forms of interaction) occurs between organisms and the physical environment; between organisms and local conspecifics; and, significantly, between organisms and other nonconspecific organisms. Ecosystems are inherently cross-genealogical—hence their characteristic absence in all but peripheral evolutionary discourse. Evolutionary theory has traditionally focused almost completely on the genealogical products of evolution: species and monophyletic taxa of higher rank. In so restricting itself, evolutionary theory almost totally ignores large-scale biological entities that form as a direct consequence of the organismic adaptations devoted to economic processes. Yet, paradoxically, evolutionary theory retains as its central concern the explanation for the origin, maintenance, and modification of economic adaptations (cf. Dawkins 1986).

We have omitted *local populations, demes,* and other partially synonymous terms from the brief list of organismic aggregates. Cross-genealogical ecosystems are readily seen as radically different sorts of entities from species; below species on the one hand and ecosystems on the other, the growing split between economics and genealogical processes within organisms is still only partially complete at the local population level. Such populations may be seen clearly in economic terms, as they are in niche theory, where local representatives of a species are seen to play a (more or less) specifiable role within a local ecosystem. Or we may view such aggregates as demes—local representatives of a species, within which the actual processes of sexual reproduction are being carried out. Yet, to a significant degree, these two sorts of local aggregates of conspecifics may be one and the same aggregate; or they may differ significantly in composition depending upon time of year and nature of the biological process of momentary greatest significance.

We will explore these problems in greater detail in the following chapters. Suffice it to say for the moment that local aggregates represent a further, yet still incomplete, severing of the economic and reproductive activities of (especially, but by no means solely) sexual, metazoan animals. Social systems, of course, fit into this level of local organismic aggregates. Thus we need to examine in greater detail the precise natures of, and interrelationships among, the various sorts of groups in which organisms find themselves—by sheer dint of their economic and reproductive activities.

•••••••••
Biotic Consequences
of Organismic Reproduction

Sex is an interesting conundrum in evolutionary biology. If, as ultra-Darwinians insist, concerted competition between organisms of like kind to maximize each one's own genetic representation in the succeeding generation is inherent in the very nature of things, reproducing by combining a mixed sample of one's own genes with an equal amount of genes from another organism hardly seems efficient. Asexual reproduction, where offspring share 100 percent of their genes with the parent, would seem by far the better strategy.

Nor, selectionists argue (to generally good effect), will it do to put the prevalence of sex off to some putative "good of the species." It may seem logical to assume that evolution will maximize the probability of survival of a species, but as George Williams (1966) was at great pains to argue, it is difficult to imagine *how* evolution can effect such a goal: natural selection is a matter of differential reproductive success—among organisms, on a generation-by-generation basis. Selection can be expected to maximize *organismic* benefits—which may have implications for long-term species survival.[1] But we cannot expect organismic natural selection to have "eyes for the future," anticipating what will await future generations. Rather, selection is a matter of retaining what worked best for the *preceding* generation. Natural selection is undoubtedly a very powerful mechanism, but species longevity is beyond its reach.

Nor, for similar reasons, will it do to argue that sex enhances variability (through the processes of recombination in meiosis and the merging of two parental sets of genes). Geneticists are sometimes prone to argue that because natural selection works on the variation expressed each generation, evolution will favor the production of variation. But once again, such an argument falters when the actual mode of action of selection is considered: variation is useful for *future* evolution—if, for example, the environment changes. But variation at any given moment is often considered a burden, a "load," that diffuses the focus of fit of a population to its "adaptive peak," to use a metaphor common in population genetics. Consequently, selection should be acting to *reduce* variation—or, more accurately, because the very fittest are by definition most favored, variation should be routinely minimized as a consequence of ongoing natural selection.

Thus modern evolutionists have felt obliged to explain how sex benefits organisms directly—despite the fact that sexual reproduction puts them 50 percent behind the eight ball in achieving their putative goal of maximizing their individual genetic representation in succeeding generations. The literature on the subject, still growing but having reached its zenith of activity perhaps in the 1970s, is impressive. Noteworthy especially are the single-authored works of Ghiselin (1974b), Williams, (1975), Maynard Smith (1978), and Bell (1982), as well as most recently, the multiauthored volume edited by Michod and Levin (1988), and, of course, many papers in the technical literature. Explanations for why there is sex at all are many and diverse—and quite possibly some or even all (rather than just one of many) of these ideas are correct. Bernstein, Hopf, and Michod (1988) have argued compellingly, for example, that sexual reproduction is the basis of diploidy—where all but occasional specialized cells (the latter including, of course, gametes) have two sets of chromosomes. It is the somatic, economic consequences of this reproductive strategy, ironically enough, that explain the prevalence of sex. Diploidy, in their view, presents an economic advantage to organisms because it facilitates repair of damaged genetic material. Not only is there a second, "spare" copy of a gene available in the event the other is dysfunctional, but the damaged allele can in some cases actually be repaired through the process of gene conversion, in which the functional allele imposes its form on its dysfunctional homologue. Increased somatic viability would easily compensate for the 50-percent loss of efficiency in genetic transmission—especially because all members of a population face precisely the same 50 percent "cost" of sexual reproduction.

But we wish to pursue a somewhat different strand of thought. Michod and his colleagues adduced an argument on the benefits, hence the prevalence, of sex based on a direct consequence of sex, namely, diploidy. There is at least another simple consequence of sexual reproduction, and it too seems to hold promise as a further explanation for the prevalence of sex—though in this instance, not through any putative direct benefit to organisms; thus it is definitely not a phenomenon encompassed by natural selection. This second consequence, in a sense, returns to the biotic perspective of earlier arguments that spoke of sex as being for the good of the species. But rather than seeing species, and other biotic systems, as *benefiting* from organismic sexual reproduction, we simply note that these systems owe their very existence to that activity.

Sexual reproduction implies a community of organisms, minimally two, but in practice ranging upwards from several to millions. Sexual reproduction, in fact, creates groups of interbreeding organisms. And there are aspects to the nature, organization, and spatiotemporal distributions of such groups that in turn have implications for the issue of the prevalence of sexual over asexual modes of reproduction: groups of sexually reproducing organisms are, by their nature, far more stable entities than are the clones formed through asexual reproduction. In other words, sexual reproduction leads to the formation of systems, or entities, with their own properties and characteristic spatiotemporal dimensions. We argue, in short, that such entities are inherently stable, and because they are stable, a feedback loop is established: the larger-scale systems formed by reproduction, being themselves stable, account at least in part for the persistence and prevalence of the underlying causal mechanism that creates such entities in the first place—namely, sexual reproduction.

Such a proposition demands a hierarchical perspective—or at least a willingness to view organisms potentially as parts of larger wholes. In our view, sexual reproduction automatically underlies the establishment of a series of nested levels of progressively larger-scale (i.e., spatiotemporally, but also in terms of numbers of component organisms) biotic systems, to which we now turn. We will begin at the level of species—larger-scale than local populations of reproducing organisms (demes—to which we return)—but central and absolutely crucial to evolutionary discourse ever since Darwin.

Species

Hull (1980) has maintained that progress in evolutionary biology— specifically in the form of useful generalizations or "law-like statements"—will depend in large measure on the willingness of biologists to abandon what he calls *commonsense ontology*. Such generalizations as a rule involve classes of events affecting kinds of entities— as when the gas laws yield a statistical generalization of the behavior of a collection of gas molecules under certain specified conditions. The problem with species, in Hull's view (and following a line of argument developed first by Ghiselin [e.g., 1974a]) is that each one is, in commonsense ontology, considered a general "kind," or class, that is, a collection of entities (organisms) of like kind. While this seems to open the possibility that each species offers lawlike generalizations, there seem to be no generalizations true of *all* species. Yet as Ghiselin (1987) has pointed out, "allopatric speciation" is in fact such a generalization (pertaining to the characteristic mode of origin of species). Ghiselin (1974a) concluded that species are, in fact, "individuals" rather than "classes"—a terminological distinction that, as we discussed in chapter 2, has run afoul of philosophical analysis. But its original *biological* import and intent are unambiguous: species are real things, not just categories of classificatory convenience. They have proper names, and, perhaps more importantly, they have beginnings (via "speciation"), histories, and ultimately terminations—extinction has already claimed the vast majority of species that have ever existed. Species are *historical entities:* they are "real."

Ghiselin (1974a) and Hull (1976, 1978, 1980) were concerned that species be labeled individuals primarily because, through time, lineages of sexually reproducing organisms are expected to change— that is, at least some of the phenotypic and underlying genotypic properties of the component organisms are expected to become modified through genetic drift and natural selection, given the mere passage of time. If such lineages—species—are true classes, they argued, they must have immutable defining properties. Instead, organismic properties are modified, and there are no eternal, unchanging, defining properties of species; therefore, species must be analogous to organisms, which remain individuals no matter how much they change during a lifetime.

The standard view of species under the modern synthesis was to see species as discrete reproductive communities, distinguishable (on the basis of organismic phenotypic properties) from other entities of like kind; indeed, Mayr (1942:152) likened species to individual organisms in his discussion of the practical difficulties faced by systematists grappling with problems of classification in closely related, geographically disjunct (allopatric) populations. It can be difficult, he wrote, to decide if two such separated populations represent two closely related species or are parts of a single, geographically variable species. But from the vantage point of an evolutionist concerned with causality, such cases are only to be expected: "If we look at a large number of such arrays (that is, species), it is only natural that we should find a few that are just going through this process of breaking up. This does not invalidate the reality of these arrays; just as the *Paramecium* individual is a perfectly real and objective concept, we find in most cultures some individuals that are either conjugating or dividing." Species, to Mayr (and to Dobzhansky [1937] as well) were best considered, at any one time, to be "real" entities.

Not so, however, through time. Mayr (1942:153), immediately after the passage in which he sees species as analogues of individual organisms, repeats the standard viewpoint on species through time that has predominated since the age of Darwin: through time, species are bound to change—*thus species have no temporal persistence or dimension.* The species concept simply does not apply temporally in the synthesis—a viewpoint upheld vigorously in the recent literature, explicitly so by Bock (1986). Simpson (1953, 1961) was especially perplexed by such a curious construct: how could the biological species concept articulated by Dobzhansky (1937) and then Mayr (1942) possibly be ahistorical, thus virtually nonevolutionary? Simpson, of course, was a paleontologist; in *his* vision, species were lineages evolving through time, each with its "own separate evolutionary role and tendencies"—a conceptualization not unlike that of Wiley (1978) and Ghiselin and Hull when they concluded that such changing lineages are best construed as individuals.

Secondary-school texts still tend to give, as a first definition of (biological) species, the notion of different *kinds* of organisms. There are groups of similar organisms in nature; allowing for differences between the sexes and for other forms of phenotypic variation, organisms within these groups are, in general, more obviously similar to one another than any of them are to organisms sorted into other groups. The second definition goes on to specify that, among sexually reproducing organisms, reproduction goes on among members within each group but not, as the overwhelming rule, with organisms from other such groups. Indeed, even the authors of Genesis remarked that like begets like; there has always been a notion of reproductive community and continuity associated with concepts of species, however much the initial emphasis might be on phenotypic similarity in the definition and recognition of species— "kinds" of organisms.

Dobzhansky (1937) and Mayr (1942) simply reversed the order, hence the emphasis, of the two sorts of conceptualizations that have always been intertwined in species concepts. That is, they saw species as reproductive communities whose members tend to resemble each other more closely than they do member of other such reproductive communities—because, of course, they share a reservoir of genetic information because of those very reproductive activities.

Why these two biologists chose to emphasize reproductive communality over phenotypic resemblance as the primary elements of their definition of species appears to reflect their predominant concern with evolutionary causality: both Dobzhansky (1937, 1941) and Mayr (1942) saw discontinuity as a theme of equal importance to the generation of (phenotypic) organismic diversity (the original and still central problem requiring explanation in evolutionary biology). They realized, or at least maintained the traditional argument, that natural selection tends to create an unbroken spectrum of intergradational variation between organisms. Yet nature seems to be broken up into a vast series of discontinuous arrays (species). Unless extinction of intermediates is adduced to explain all the discontinuity, some additional mechanism needs to be invoked. Dobzhansky, in particular, believed that discontinuity enabled each species to focus more closely on its own adaptive peak. He believed that discontinuity is useful and will be encouraged by natural selection. Both Dobzhansky and Mayr developed arguments that saw reproductive isolation, developed as a consequence of genetic divergence in geographic isolation, as the forerunner of general phenotypic discontinuity between species. Their theories differed in detail and have since been modified by themselves and many other evolutionary biologists; these issues are reviewed in greater detail in Eldredge 1989 (ch. 4).

The point to be stressed here is the interplay and tension between conceptualizations of species as either real entities or as simply collections of similar organisms and the related, but distinct, debate over whether species are to be conceived primarily as interbreeding communities (whose component organisms resemble each other closely) or as groups of similar organisms that happen to interbreed with one another and not with members of other such groups. There is a tendency (by no means universal) to consider species as real entities (individuals) if they are seen to be lineages—that is, historical, temporal, reproductive communities with ongoing (bi)parental ancestry and descent; even the founders of the biological species concept saw species as real entities at any one time. However, when the emphasis is on aspects of general organismic phenotypes—whether species are seen simply as groups of similar organisms or some special aspect of similarity is singled out for group definition and designation (as with cladistic recognition of genealogical structure *within* biological species; see below)—there is a strong sense that species are at most loosely knit communities and at least simply collections of similar organisms. If species are not wholly groups of similar organisms arbitrarily designated by some taxonomist (a view still upheld by some biologists, though by few evolutionists), nonetheless species may be nothing more than a loosely knit collection of similar organisms.

This last view is actually the version of Hull's commonsense ontology that predominates in evolutionary biology: under this view, species are not true classes, simply arbitrary collections of similar organisms. But neither are they complex systems that can usefully be seen to have organization, properties, functions, or even discrete origins, histories, or terminations (beyond the trivial final demise of the last member of any given lineage). All biological dynamics, be they economic, reproductive, or evolutionary (arising, that is, at the interface of economic and reproductive activity—as through natural selection), are to be understood solely with reference to organisms. To maintain that species, however conceived in detail, are themselves to be construed as entities is to indulge in reification.

And why not look at the biological world strictly in these terms? Organisms patently exist, natural selection is a preeminent causal process among organisms within populations (which need have no properties beyond those of the component organisms), and, in short, there seems little reason, especially when we confine our gaze to a single time slice (in particular, the Recent), to complicate matters by seeing large-scale collections of organisms as true entities, with structures, properties, and functions. Evolutionary theory has gotten along fine (many would say) without the excess conceptual baggage that stems from seeing species (or demes, monophyletic taxa, and ecosystems, for that matter) as real entities.

Ontological considerations of nature arise, so to speak, through themselves: we seek the most accurate description of nature possible. If organisms are parts of larger-scale biotic systems, and if such systems exist in their own right, we quite naturally need to know that. And we need to characterize those larger-scale entities carefully. But, in practice, there has to be a more practical, compelling reason to adopt an ontological stance as unusual (with respect to tradition) as the notion that species and other sorts of biotic entities actually exist and pose implications for our understanding of the nature of the evolutionary process. We need to show that there are compelling reasons to believe that organism-centered ontology (or an ontology in which organisms stand as the largest entity, composed of crucial elements, particularly genes) is an incomplete description of biological nature; and *that* is accomplished primarily by the demonstration that there are phenomena unexplained (or insufficiently explained) without an expansion of biological ontology to embrace larger-scale entities.

However species are specifically defined (and, for the moment, it makes little difference), the recently reemphasized phenomenon of *stasis* in the fossil record has had a direct bearing on evaluation of the ontological status of species. Stasis refers to the tendency of the preserved aspects of the phenotype of organisms to remain relatively stable within a lineage through the vast bulk of the documented stratigraphic range of a species lineage. "Relatively stable" allows for within- and among-population (i.e., geographic) variation, but such variation is usually far more oscillatory, clinging around some nearly stable mean rather than moving directionally, and especially linearly, over long periods of time.

Such relative phenotypic stability was known to Darwin's paleontological contemporaries; indeed, Darwin himself alluded to it by his sixth edition of *On the Origin of Species*. The term *stasis* was coined for this phenomenon (Eldredge and Gould 1972) in conjunction with its resurrection as an issue to be taken seriously. It is now clear that phenotypic change is not the inevitable consequence of the mere passage of time that evolutionary biologists ever since Darwin have more or less always assumed—the very attitude that led Mayr (1942:153) to suppose that species inevitably evolve themselves out of existence because organismic phenotypes inevitably will change over time.[2]

Levinton (1988) has argued that stasis is not an attribute of species but simply lack of accumulation of concerted organismic change through time; stasis is a cumulative property of organisms and not a property of species per se. However (obviously) true this contention is, a practical effect of stasis is that the fossil record has many more examples of discrete packages of phenotypic diversity—in which groups are phenotypically disjunct at any one time (the observation agreed upon by virtually all biologists) and *over time* as well. Empirically, a far greater percentage, easily a preponderance, of lineages known from the fossil record are distributed as spatiotemporally discrete packages of phenotypic properties. It is understandable that paleontologists have labeled such packages "species" over the years.

Further, the composition of monophyletic taxa (i.e., taxa of rank higher than species) undergoes change through time: new species appear, while others disappear. In particular, Eldredge and Gould (1972) noted that stasis effectively removes the traditional explanation for long-term evolutionary trends: that such trends are merely large-scale manifestations of microevolutionary change, generally if not exclusively under the control of directional natural selection. But there is a paradox latent in the recognition of stasis. If species fail to change directionally through time (especially in the direction of the long-term trend), and especially if the ranges of ancestral and descendant species overlap, the actual pattern of a directional trend might, and arguably *must*, arise from some process of *sorting* (Vrba and Eldredge 1984; Vrba and Gould 1986; Vrba 1989; see Eldredge 1985 and 1989 for discussion) of the component species within the monophyletic clade. In other words, there may be one or more causal processes biasing species composition within a clade that accounts for such macroevolutionary patterns as trends, adaptive radiations, and the like. It is this empirical phenomenon of changes in composition of spatiotemporally discrete (phenotypically) lineage segments within monophyletic clades that has prompted some investigators to take seriously the possibility that species are spatiotemporally bounded historical entities that have direct roles to play in the evolutionary process.

Species as Entities: What Are They and What Do They Do?

Perhaps the crucial difference between seeing species as individuals or as classes lies in the arena of interactions and functions—the interrelations between parts that lends coherence and cohesion to the whole, and the related issue of what function(s), if any, the whole (i.e., the entire species) plays in nature. Whereas the members of sets, categories, or classes certainly may interact in various ways, such interaction is usually not a criterion for membership. Thus, in biology, an organism belongs to a taxon (including species so conceived) if it possesses the requisite phenotypic properties by which the taxon is defined and recognized.

That interaction among parts is important, even crucial, to a claim of individuality of an entity is clearly seen in models of manufactured items. No one would claim that the Edsel constitutes an individual. Yet Ford manufactured the Edsel model for only a few years in the 1950s; the Edsel was certainly spatiotemporally bounded. Moreover, the design concept changed slightly over its brief history—not all Edsels were alike from year to year—just as species change (evolve), a prime consideration for Ghiselin (1974a) and Hull's (1976ff.) first arguments in behalf of species individuality. True, Edsels could be manufactured anywhere (i.e., there was no single point of origin), and production could conceivably cease for years, to be resumed at some later date (as in fact happened when World War II temporarily halted production of many manufactured items). Nonetheless, Edsels and all other models of manufactured items can reasonably be construed as spatiotemporally bounded but not, really, as entities. Each example (i.e., each car) is a specimen, a token, of the Edsel type. Each car is held together by the cohesion of its molecular structure and the imposed joints between parts. There simply is no such cohesion between individual cars— the main reason the Edsel is no individual, system, or discrete entity. That species are not Edsels, Mustangs, and LTDs arises from their internal structures.

The biological species concept definitely invokes a particular sort of interaction among component organisms as a source of cohesion of species. That interaction, of course, is sexual reproduction—the act of putting to use those organismic phenotypic traits (including anatomical, physiological, and behavioral traits) that form the set of reproductive adaptations for each organism. Species are reproductive communities, cohered by a plexus of ongoing parental ancestry and descent. And even though panmixia seldom if ever prevails, the general assumption is that enough matings between populations occurs for some degree of genetic *mixis*. The recent apparently explosive spread of the P4 element (a novel segment of DNA) throughout the world's populations of *Drosophila melanogaster* is apparently a rather graphic case in point.

There is no comparable cohesion stemming from general use of the nonreproductive aspects of organismic phenotypic attributes— beyond, that is, the multifaceted interactions that occur among organisms of like kind on a purely local level. The distinction here is crucial: that juncos constantly scrap and jockey for position in grain-rich winter fields is certainly a form of interaction that arguably lends a form of cohesion to the localized population. As a first approximation, the behavior is purely economic: there is competition for food, and the juncos are putting into effect their complex behaviors and anatomies (in short, their adaptations) pertaining to feeding. (That much of the male-male and male-female jockeying reflects mating behavior destined to occur elsewhere in later months must also be acknowledged.) But the point here is simply that local economic interactions between organisms of like kind do not lend the same real or potential coherence to other such populations (e.g., other populations of dark-eyed juncos) elsewhere.

Local reproductive interaction can thus be seen as part of a larger-scale plexus of interaction—unifying a species interregionally as well as through time. Local economic interaction, however, is held to be of purely local significance. Both aspects of this generalization require further analysis before we can be confident that they are apt descriptors of the nature of species. We need to look a bit more closely at the problem of evident nonreproductive continuity in those many situations where single species are distributed widely but in disjunct (sometimes starkly so) populations. And we need to ask: if economic interactions do not unify large-scale (spatiotemporally) entities we can call species, why is there such a marked degree of similar economic behavior among most geographically (and stratigraphically) separated populations of conspecifics? The reason evolutionary biologists historically have had little difficulty imagining species (and even higher taxa) to be in some sense ecological entities is that members of taxa seem to share ecological attributes so closely.

Reproductive Cohesion in Space and Time

Sewall Wright is best remembered as a mathematical geneticist given to theoretical population genetics, from which his evolutionary theory emanated, and to analysis of experimental breeding data. Wright, more than anyone else, is responsible for the most generally accepted vision of the characteristic nature of internal structure and spatial distributions of species. Using the term *colonies*, Wright (e.g., 1931, 1932) painted the graphic picture of species distributed discontinuously in a complex array of semi-isolated populations. Indeed, Wright's "shifting balance" theory is based precisely on such structure: each such semi-isolated colony (later termed *deme*) was postulated to face, through chance and selection, somewhat different evolutionary histories. Evolution within a species is a summation of within- and among-deme effects.

Dobzhansky (1937) was the earliest and most forceful sympathizer with Wright's evolutionary views, including that of the nature of species structure and organization. Though Wright's influence is widely considered to have waned as the synthesis was fashioned and "hardened" (e.g., Provine 1989; Gould 1980), there is little doubt that his view of species being typically composed of a number of quasi-isolated local colonies has always struck field naturalists and evolutionary theorists as an accurate description of the vast majority of all existing species. No species is known to be perfectly homogeneously distributed throughout its geographic range. Rather, species distributions reflect environmental heterogeneity: local representatives of species are found where suitable habitats exist; they, quite naturally and by definition, do not occur within their overall range where unsuitable habitats exist. There are, of course, exceptions: local populations may be absent from habitats that appear identical to occupied territory elsewhere, and they may be found in areas that appear unsuitable. There is no absolute determinism whereby habitat parameters strictly dictate if any given species will or will not be present. But none of these considerations detracts from the simple generalization that every species displays heterogeneous distributions within the known boundaries of its total geographic distribution. Indeed, long before Wright made such distributions explicit as a cornerstone of his shifting balance theory, heterogeneous distributions of species had been familiar to all field naturalists.

Such considerations undoubtedly underlay Mayr's (1942:120) inclusion of the otherwise troublesome phrase "actually *or potentially*" (italics ours) when he spoke of interbreeding as the cohesion, the very defining quality, or species. Real, actual interbreeding means a complete plexus of interbreeding uniting an entire species throughout its range on a regular basis. Potential interbreeding, on the other hand, acknowledges that species are seldom so geographically restricted that such total interaction is possible each generation.

Mayr, reacting to criticism over his use of *potential*, later dropped the term. Most criticisms focus on the operational implications of the term. If we can, at least in principle, study organisms in the field to determine patterns of actual reproductive behavior, how are we to assess empirically the existence of potential interbreeding? Yet there must be a component of "not-now-actual-but-in-principle-possible" if a notion of reproductive cohesion is to be reconciled with what everyone agrees is the heterogeneous nature of species spatial structure.

One (relevant if not wholly satisfactory) answer to this dilemma is that scientists in general are forever confounding ontological issues with epistemological considerations. Nor is this altogether bad: biologists quite naturally expect to be able to know how they can tell a species when they meet one. It is only natural that they expect to derive the criteria for recognition from what species are. Yet what species are is not the same thing at all as the methodological rules we might derive by which we may expect to know them. The entire critique of the biological species concept by Sokal and Crovello (1970) hinged precisely on the (alleged) difficulties of applying the implied criteria of species recognition rigorously and consistently to individual cases.

Yet there is another way of looking at the problem that has the happy effect of addressing ontological issues and epistemological concerns simultaneously. Biologist H.E.H. Paterson (1985 and earlier works) has formulated and promulgated the concept of the Specific Mate Recognition Systems (SMRS). Paterson's argumentative springboard has been the "relational" nature of the biological species concept (BSC). Both Paterson and the proponents of the BSC see species as reproductive communities. Paterson, however, points out that the notion *reproductively isolated* (as in the phrase "reproductively isolated from other such groups"—the final segment of Mayr's [1942:120] well-known "short version" of the BSC) in effect means that species are defined (and recognized)? sheerly by dint of their separateness from other entities of like kind.[3] Paterson, in contrast, sees species as entities existing in their own right: they arise, and have cohesion, for reasons originating in the reproductive biology of their component organisms; and the reproductive adaptations (i.e., his SMRS) exist, originated, and are put in use each generation for reasons intrinsic to the organisms within a species—and *not* because of the existence of *other* organisms, with other reproductive adaptations.

And that is precisely what Paterson's SMRS *is:* the set of organismic reproductive adaptations by whch males and females in particular (but also, of course, members of like sex) recognize each other as potential mates (or reproductive competitors). Nor should the word *recognition* be interpreted too literally, or strictly in terms of (conscious) human perception. Recognition of appropriate mates entails a vast range of chemical and sensory perception mechanisms. Mate recognition is certainly not synonymous with mate choice. It applies as well to chemical recognition of gametes shed freely into sea water or to pollen driven on the winds as it does to male and female African waterbuck recognizing the white rump circle, characteristic horn shape (of males), and behaviors that tell each other that their gametes will likewise prove compatible. Thus the botanical complaint that the notion of the SMRS is a function of zoocentric biologists misses the point utterly: the SMRS is that set of adaptive organismic features that allows appropriate mates to find and successfully mate with one another—regardless of what kind of orgamisms they are.

The immediate ontological consequence of this formulation is strikingly clear: because successful sexual reproduction depends utterly on a finely tuned, functioning set of coevolved adaptive organismic features, species are natural entities, systems of mating organisms that share a particular set of reproductive adaptations— a particular SMRS. Species, as reproductive units, exist in nature as a simple consequence of the necessity for there to be an interactive set of (reproductive) adaptations.

The epistemological consequence of this formulation is equally obvious: we define and recognize any given species by using the same criteria of recognition used by the organisms themselves. Thus, in the famous and still unfolding drama of biology's attempt tc understand leopard frogs in North America, Moore's (e.g., 1949) original story of a single, large, clinically interbreeding biological species *Rana pipiens* (where the end members were reproductively isolated) has been replaced more recently with a picture of numerous species (within two major groups) distributed for the most part parapatrically (i.e., with range boundaries adjacent but nonoverlapping (Hillis, Frost, and Wright 1983; Hillis 1988; cf. Moore 1975). The frogs all look pretty much alike but tell one another apart mainly by vocalizations—the main component of their SMRS.

Note, too, that the only requirement is the demonstration of shared reproductive adaptations: no longer is it necessary that interbreeding be observed before a biologist can reach a judgment as to species identification (though the leopard frog example illustrates how subtle SMRS differences of other species can appear to humans)—a point both Paterson (1985) and Vrba (1985) have made. Thus, even in fossils, provided a portion of the anatomy involved with the SMRS is preserved, isomorphy in structure bespeaks a shared SMRS as a direct guide to the composition of species. Gone, then, is the objection on epistemological grounds to Mayr's potential interbreeding. It is the *sharing* of reproductive adaptations, rather than the state of panmixia, that matters most when considering what species are as well as in deciding how we biologists are to recognize them.

But what of the issue of ongoing functional *interaction*—in the case of species, *reproductive* interaction—as the real source of cohesion of a species? After all, it could be argued, mere sharing of a set of reproductive adaptations, if not ever put to use between far-flung populations, more closely resembles sets of similar entities than it does a vibrant, functioning whole held together by interactions among its parts. Surely those adaptations must be put to use in network fashion—in a skein of interbreeding that does link all the far-flung, semi-isolated local colonies that so typically make up any species.

There are considerations that go beyond the simple supposition (buttressed by case examples, such as the P4 element of *D. melanogaster*) that argue that "gene flow" to some extent exerts specieswide, if not homogenizing, effects. In an influential paper a generation ago, Ehrlich and Raven (1969) made a forceful case that gene flow is not as all-pervasive as others (notably Mayr) had been arguing in their discussions of the biological species concept. Replacing gene flow came a notion of homeostasis, where substantial divergence often does not occur even in the absence of gene flow—contrary to the more conventional view that saw selection-driven diversification as the inevitable norm unless counteracted by the homogenizing influence of strong, appreciable gene flow between far-flung populations within a species.

Yet there is manifestly more to the situation than stabilizing selection keeping divergence in check. For one thing, Paterson's notion of the SMRS makes it abundantly clear that there is a natural distinction between reproductive and all other organismic attributes (many of the latter can be viewed as *economic* attributes). Traditional speciation models—certainly the one under which Ehrlich and Raven wrote—saw reproductive isolation as a simple by-product of any form of diversification (i.e., reproductive or economic) accruing until the point of (gametic, genetic) incompatibility is reached. Thus failure for speciation to occur between far-flung isolates implies, in their model, that little or no divergence of any kind occurred.

Paterson, however, makes it clear that speciation (the origin of new reproductive communities from old) is a matter solely of the development of a modified set of reproductive adaptations—in other words, a new SMRS (more correctly, a distinct SMRS, modified from some preexisting condition). Failure for this to occur without reproductive contact between two populations requires a theory of stability only of reproductive adaptations, not of all organismic features. That this is intuitively grasped by many systematists seems to be borne out by patterns of revision in systematics of North American bird species of recent years: well-differentiated taxa formerly considered distinct, albeit closely related, species (e.g., Audubon's and myrtle warblers, now lumped as "yellow-rumped warblers"; the various juncos now known collectively as "dark-eyed juncos") are lumped, despite their plumage and other differences, if they are known to interbreed along zones of contact, and if their songs cannot be readily differentiated. Other birds that have occasionally hybridized have not been lumped ("synonymized," e.g., blue-winged and golden-winged warblers). Even though plumage is arguably every bit as much a part of a bird's SMRS as its song or other more obviously reproductive aspects of behavior, the point here is that it is the status of similarity of features traditionally implicated most closely in reproductive biology (e.g., song over plumage pattern) that has evidently made the difference in decisions on the status of species composition in instances where hybridization has been documented in North American birds.

But, again, can we speak of species as reproductive communities when the shared SMRSs are not actually interconnected, at any given time, throughout the far-flung systems of semi-isolated local populations within a species? A final consideration here is time. The order of magnitude of time for stability of anatomical systems generally and (to the extent it is known) of SMRSs in particular ranges from tens through hundreds of thousands, often millions, of years for terrestrial organisms to tens of millions of years for marine organisms (especially invertebrates) (Stanley 1985).

Wright (1931, 1932) imagined that his semiautonomous colonies, subdivisions of species, would not only have partially independent, and therefore different, evolutionary histories from each other. They would also tend to be ephemeral. Local populations are constantly being established, becoming extinct, and merging with one another. Wright was, understandably, concerned with the fate of genetic variation under such a kaleidoscopically changing structure. Indeed, it was the ephemeral nature of local populations that led Darwin (1871) to pronounce species as *"permanent* varieties" (italics ours). When new SMRS are established, the opportunity to swamp (and perhaps lose) genetic differentiation through merging of populations, is lost; and, as we shall see, largely by virtue of their typical structure, species are difficult to drive to complete extinction. Thus phylogenetic novelty is effectively injected into the "phylogenetic stream" through speciation (see Eldredge 1989, chapter 5 for more discussion of this point)—a fact that Darwin clearly saw.

The implication for the present discussion is that the structure—and degree of interaction between local populations—that biologists can observe at any one moment is characteristically ephemeral, given the expected and characteristic duration of the species as a whole. Species exist on a temporal scale that transcends, by orders of magnitude, the life spans of their constituent organisms. Organisms with relatively short life spans and generation times (that is, relative to human scales of existence and observation) may be expected to obviate this problem of understanding the degree to which reproductive interactions *do* obtain between far-flung populations. This is precisely the relevance of the P4 element example of *D. melanogaster*. In the fullness of time, not only might there be eventual genetic connections (i.e., even if, at any one time, there seems to be functional reproductive isolation of the local populations of some species from the rest of the species), but those same far-flung populations will in all likelihood not persist: they will become extinct; or they will regain contact—and either merge completely, or simply begin (again) to share genetic information; or they might develop their separate SMRS—and become forever separated. This long-term perspective effectively counters objections to seeing species as arising and being maintained as a direct result of reproductive interactions among their constituent (sexual) organisms.

Species and Economics

Reproductive interaction among sexual organisms creates local reproductive populations—demes. Demes may remain, at least for some time, isolated from other conspecific demes; yet it is their fate to interact, to have some degree of reproductive (genetic) contact, up to the point at which they may merge completely. Thus is a species bound together by shared possession and (eventual, if sporadic) *use* of reproductive adaptations, the SMRS unique to each species. We now must consider whether comparable use of economic adaptations likewise leads to associations of like organisms spread discontinuously over a geographic range, or whether use of economic organismic properties leads to the formation of rather markedly different sorts of groups altogether.

We examine the biotic consequences of economic behavior in more formal and greater detail in the following chapter. We restrict our commentary here to two related issues pertaining specifically to species: does *use* of economic attributes by organisms yield entities that can be confused with any of the traditional conceptualizations of species? And is there any sense in which species per se can be usefully and accurately viewed as economic entities? In particular, can species be construed as both economic and reproductive entities?

The preponderance of organismic phenotypic features pertains to economic functions. Naturally, if species are construed merely as collections of similar organisms, most of the similarities used to distinguish one species from another—to sort specimens into piles said to represent different species—will be economic rather than reproductive. We shall deal further with the issue of species recognition based on different sets of organismic attributes. But the issue at the moment is cohesion of entities we might call species arising from interaction among putative conspecifics—interaction, moreover, that arises from the uses to which those two separate categories of attribute (economic and reproductive) are put. In a nutshell, does economic interaction *among organisms of like kind* lead to regional—as opposed to purely local—aggregates that we might call species? The issue here is cohesion: everyone agrees that giraffes in Kenya look and behave much as do giraffes in South Africa—and that most of that similarity in appearance and behavior involves somatic features involved with economic existence rather than reproductive activity. But is *Giraffa camelopardalis* coherent in any meaningful sense because the nature of the economic existence is in some extended sense interactive between Kenyan and South African populations—the way, for example, we have argued reproductive interaction in an extended sense does pertain to giraffes (and all other far-flung species)?

The answer must be "no." The fundamental reason, dealt with more extensively in the next chapter, is that use of economic attributes may indeed lead to associations of organisms of like kind (and reproductively compatible) on a local level, but these local economic entities in turn interact with other such entities (populations, or avatars; Damuth 1985). It is the nature of economic behavior to cut across the grain of genealogical kinship and (more to the point) reproductive community. Indeed, economic behavior is mostly a matter of energy resource use and exploitation—by its very nature carried on outside the reproductive community. Like beget like; but they eat unlike. Economic existence is a moment-by-moment affair, and the economic life of any organism only impinges on others—of like kind and not—with which it has contact. Thus the standard picture of ecology: economic interaction leads to local economic populations (avatars) that have niches (play roles) in local ecosystems; regionally, local ecosystems interact (by matter-energy transfer), thus larger-scale systems are seen to cohere (as when a barred owl leaves its woodland perch to take a vole in a farmer's field). The interactions between microecosystems to form large-scale systems is characteristically, if not invariably, cross-genealogical. There seems to be no promise of regionally-embracing patterns of organism-organism interaction arising from economic activity that would lead to the cohesion of any sort of system we would want to identify as species.

Thus, to return to Wright's standard picture of species deployed in semiautonomous local colonies: at any one moment, the reproductive and the economic activities of local organisms are irrelevant to the lives of conspecifics in other such localized colonies; but reproductive interaction does spread, and the sharing of reproductive adaptations has functional, interactive consequences that realistically bind a species into a loosely organized but functioning whole. Sharing of economic adaptations offers no comparable interactive device to hold a species together. Introduction of conspecific organisms from foreign colonies may alter the complexion of economic lives of a local population, but the impact is ephemeral unless, of course, the genetic basis of differences between foreigners and locals is injected into the local population.

And that is the crux of the matter. Organisms from far-flung colonies resemble each other—be it in reproductive or economic features—simply because they share genetic information. They share that information because they inherited it from a common ancestor and because of actual and latent continued sharing of that genetic information. It is an irony, but only an accident, that conspecifics share most of their resemblances in economic rather than reproductive features. They are economically similar because most specifiable organismic features are economic (somatic) rather than reproductive—and hence most genetic information pertains to such features. *But it is the fact that they share the information and still exchange it that unifies the species;* it is not simply their similarity, nor any imagined cohesion emanating from actual use of economic attributes, that unifies these things we call species. As far as economic organismic attributes are concerned, species really are mere sets of similar organisms.

As a simple extension, it follows that species as whole entities cannot be economic entities—that is, cannot be seen to play a role in some imagined large-scale ecosystem. This is not a trivial point; much of macroevolutionary theory is devoted to seeing species (and larger-scale taxa) as possessing economic roles in nature comparable to the niches of local populations (see Eldredge 1989). That lions everywhere behave similarly (economically and reproductively) cannot be denied (although many subtle regional differences occur); but it cannot consequently be maintained that lions as a single, united species play a concerted economic role in some far-flung (southern Africa-India) ecosystem. If species were to be construed as parts of large-scale regional economic systems, it would have to be supposed that the regional boundaries of many species would routinely be exactly coincident. In particular, this would imply that the current geographic boundary of every species is coincident with the boundaries of a large-scale ecosystem. Historical biotas do display integrity and regularity in distribution, such that the distributions of different species roughly coincide to degrees only rather recently (re)appreciated by biogeographers; nonetheless, the typically overlapping geographic ranges of species preclude recognition of coherent, regional economic systems of which entire species can be said to be parts. Regional ecosystems cohere in ever-larger regional systems, with organisms of different species (often playing analogous ecological roles) in different parts. Regional economic systems bear no necessary relation to individual genealogical systems (species, let alone higher taxa); any coincidence between the two is purely that.

Yet cats will be cats. Not only do all lions, from the Kalahari to India, look and behave rather alike, but all cats (or at least *most* cats) display almost stereotypically similar morphologies and behaviors—as in their hunting activities. In the metaphor of the synthesis, derived from Wright's original usage of the adaptive peaks on "adaptive landscapes," this is because all cats share a singular "adaptive zone" or an "adaptive range" of closely similar adaptive peaks. The metaphor implies functional, economic cohesion—and concerted, economic role playing in large-scale economic systems. What is really at stake, of course, is the shared distribution (taxon-wide, whether the taxon be a single species or a larger monophyletic taxon) of economic adaptive traits of organisms. Organismic features, whether or not they are construed as adaptations, can be retained in more or less unmodified form—or further modified—in the course of the phylogenetic history of taxa. That taxa are defined and recognized (or simply characterized) by sets of similar organismic features is no reason to conclude that the taxon itself is an economic entity.

There has been, in other words, a strong tendency to confuse taxon-wide distributions of organismic properties with the functional implications of those properties—including the supposition that taxon-wide distribution of economic traits implies that a *taxon* utilizes those traits and is thereby an economic entity unto itself. As we shall see, species are further to be taken as genealogical entities because they do reproduce—make more entities of like kind. They are not genealogical entities because their component organisms reproduce, but rather because there is a process—speciation—that entails the origin of descendant from ancestral species. As we have seen, there is no reason to think that there are comparable *taxon-level* economic properties.

More Epistemology: Economic Versus Reproductive Traits in Species

There is another aspect to the tension between species-as-reproductively-cohered entities versus species-as-economically-cohered entities (or at least as sets of economically similar organisms): the two sets of features are often, perhaps even usually, out of phase. In other words, organismic attributes pertaining to reproduction are often distributed more, or less, widely than shared economic traits at the species level loosely construed. This fact in itself is informative, since it bolsters the conclusion that, whatever species may be, they cannot in most cases be expected to be both economically and reproductively *defined*, and still less both sorts of entities in a functional sense.

Phylogenetic systematists (cladists) generally prefer species to be treated like any other taxon category; that is (and quite apart from any issue of species as classes or individuals), species ought to be recognized by joint possession of shared evolutionary novelties unique to that particular taxon. The point is tricky, not least because species are widely and conventionally held (i.e., in standard evolutionary theory) as ancestral to other species. Thus no species ancestral to any other species can possess unique evolutionary specializations more advanced than its descendant, as a matter of definition. Among cladists sensitive to this point, however, is the thoroughly legitimate desire to see the principles of genealogical precision in systematics applied as thoroughly and consistently as possible. The problem arises only when economic differentiation is more finely developed than is reproductive differentiation—that is, at and within the species level.

Botanists have routinely opposed the BSC since its early promulgation in the late 1930s. That is because many perfectly good botanical species routinely hybridize freely, with no apparent diminished reproductive capacity of the offspring. In such instances, clear-cut morphological (economic) differentiation has outstripped reproductive differentiation. Cladists in general would agree: particularly when there is solid evidence that one portion of a reproductive community (biological species) is more closely related to some other separate reproductive community—because the descendant species and a differentiated portion of the putative ancestral species share evolutionary novelties not found elsewhere within the ancestral species—the argument is to break up (conceptually!) the ancestral spe-

FIGURE 4.1.

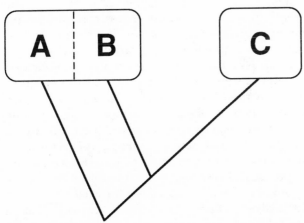

The collision between SMRS-defined and genealogically
defined "species." Here, A + B share an SMRS, that is,
constitute a "biological species." Synapomorphies (of
non-SMRS, i.e., economic, features) show that B is genealogically more closely related to species C (with its
own SMRS) than to A—suggesting to some that A and
B should be regarded as separate species. For fuller
explanation, see text.

cies into two (or even more, if the evidence warrants) differentiated
species (fig. 4.1). The argument is genealogical in that it is based on
the distributions of shared evolutionary novelties, consistent with
genealogical practice at all systematic levels. Ironically, the argument is agenealogical in that economic rather than reproductive
attributes are given sway in taxonomic practice. The argument makes
sense, particularly as an attempt to make all of systematics more
internally consistent. From the standpoint purely of elucidating
phylogenetic history, it of course makes no difference whether reproductive or economic attributes are used in the search for synapomorphy. After all, three middle-ear bones (two newly added over
the primitive amniote condition) *and* placentation are hallmarks of
placental mammals.

   It is only when we seek to characterize the organization of nature
from the ground up, as the product of interactions among component parts, that a conflict emerges. There are coherent entities
formed through the use of (sexual) reproductive adaptations. These

have been called species. There are groups of similar organisms whose similarities may be in the main economic. These have also been called species. Often the two more or less coincide, simply because reproduction ensures a degree of ongoing genetic similarity—which, of course, yields phenotypic similarity. There is, moreover, little reason to suppose that organismic economic interactions lead to large-scale entities we might call species. The only residual conflict, then, resides in the different research programs of systematists—historians of genetic information (though not all see themselves as such)—and functionalists who are concerned with causal interactions among parts of biotic nature. When economic information is more finely differentiated than reproductive adaptations, and only then, a conflict will arise over what will be called species. The conflict is legitimate; once it is understood for what it is, we can continue with our analysis, which is predicated on a functionalist rationale that natural entities arise in biotic nature as a simple consequence of the use of the two different sets (economic and reproductive) of organismic attributes. For species, only the reproductive set has interactive consequences for the origin and maintenance of entities we wish to call species.

Species as Genealogical Entities: Speciation and Monophyletic Taxa

Species are reproductive communities unified by organismic (sexually) reproductive functions. Those functions arise from behavioral integration of organismic reproductive and physiological properties. The reproductive attributes of males and females of any particular species separately are adaptations, subject to stochastic and deterministic (i.e., selection) processes governing both stability and modification. Together, the male/female reproductive adaptations constitute the SMRS of the species. It can be argued (cf. Eldredge 1989, ch. 5) that the SMRS is the only truly emergent species-level property arising from intrinsic organismic properites. (Species spatiotemporal distributions are another set of emergent properties, arising from extrinsic organismic features.)

    If male and female reproductive attributes and functions are adaptations, forged into a species-specific SMRS by a form of co-evolution, we need to ask what form of deterministic causal process shapes, maintains, and further modifies such adaptations. And it follows directly from earlier discussion of the nature of natural and sexual selection (based on simple recognition of a dichotomy of classes of organismic attribute) that *sexual* selection is the particular form of organismic selection effecting stasis and change of organismic reproductive properties.[4]

This is a rather novel usage for the notion of sexual selection. The relatively recent resurrection of sexual selection as a legitimate area of concern in evolutionary biology has focused almost wholly on within-population variation in sexually related morphology and behavior leading to within-sex differential reproductive success. Yet it is clear that sexual selection is involved at the higher level of species as well: every instance whereby organisms recognize one another (in the most general sense discussed above) and successful mating ensues is an instance of "stabilizing sexual selection": the SMRS functions by and large conservatively. Organisms bearing the typical species-specific SMRS are in all probability the ones most likely to succeed at reproduction. Male songbirds with markedly deviant songs are far less likely to attract females and successfully mate than those whose songs are closer to the standard for any particular species.

If we can readily imagine stabilizing sexual selection arising from normal use of the SMRS, we need to consider as well the circumstances under which the SMRS can and will be modified under a form of "directional sexual selection." Paterson (e.g., 1985) has considered this problem, often in conjunction with more general changes in selective regimes pertaining as well to economic adaptive modifications arising under natural selection from habitat changes for a portion of an ancestral species. For example, Paterson (1985) cites an example of an African frog species whose call differs markedly from that of its closest known relative; the abrupt change in SMRS of the putatively derived species is attributed directly to assumption of life in pools in the vicinity of waterfalls—which, in this scenario, pose for mating calls an acoustical problem not encountered by the ancestral species.

Somewhat less dramatically, it seems plausible that both drift and selection reflecting diverse conditions under which the SMRS is operating will lead to diversification—geographically based variation—in SMRS characteristics within a species. Ryan and Wilczynski (1988) present just such a case for species of cricket frogs in the southwestern United States. Fragmentation of species is an obvious consequence of significant SMRS diversification within ancestral species—in a manner generally consistent with (though focused primarily directly on reproductive attributes) the allopatric model that has remained the preferred theory of speciation since the late 1930s.

In other words, new reproductive communities occasionally spring from old through SMRS changes—changes that are readily attributable to drift and sexual selection, generally allopatrically. Species, then, make more of themselves—though the process is generally seen as one of fission, and therefore topologically analogous to asexual, rather than sexual, reproduction.

There are thus, as an automatic consequence, strings of genealogically connected, ancestral and descendant, species. Such skeins of genealogically interconnected species are, of course, the (monophyletic) *taxa* that form the subject area of systematic biology. The Linnaean hierarchy recognizes levels of relatedness among such skeins of species; for present purposes, however, rank is immaterial—any string of phylogenetically interconnected species constitutes a monophyletic taxon. We need to briefly consider the properties of these large-scale genealogical entities: how are they unified? What do they do—and, in particular, do monophyletic taxa make more of themselves (thus forming parts of some still larger-scale system), and are they in any sense economic entities?

We can conclude, at once, that because monophyletic taxa are the outcome of reproductive activities (ultimately of organisms, leading up through the fissioning of species), they are themselves inherently genealogical entities. Their internal coherence and their persistence through time depend upon the fates of their component parts: species. In particular, a monophyletic taxon arises when an ancestral species bearing the phylogenetic markers (traits, or evolutionary novelties, used to recognize any particular monophyletic taxon) arises; it ceases to exist (becomes extinct) when its last component species becomes extinct. Viewed another way, it is speciation—continued production of new species within a monophyletic taxon—that keeps a monophyletic taxon going. Cessation of speciation, coupled with extinction of component species, results in the extinction of monophyletic taxa.

Further, because monophyletic taxa are genealogical strings of species and species are construed *not* to be economic entities, it follows very simply that monophyletic taxa are likewise not economic entities. We need to examine just one more category of property: do monophyletic taxa in any sense make more of themselves, that is, reproduce? The literature of evolutionary biology is filled with analyses that take it for granted that monophyletic taxa (taxa of rank higher than species—genera, families, and so forth, as ranked in the Linnaean hierarchy) do in fact give rise to descendant taxa, as in the assertion that mammals and birds (separately) evolved from reptiles.

Yet it again follows as a simple consequence of monophyletic taxa being phylogenetically connected series of species that in a meaningful, functional sense, it is not an entire higher-ranked taxon (i.e., a group of two or more species) that, with all species acting conjointly, gives birth to some new taxon. After all, in any given case, most species within a monophyletic taxon at any given point in time are already extinct. Monophyletic taxa—and this point transcends epistemological considerations—are exactly like asexual clones. Clones cannot be distinguished (apart from distributional criteria, such as different sections of a petrie dish) unless and until heritable differences are detected between them. Exactly the same is true of monophyletic taxa. It is notorious that the groups that systematists have the most difficulty making sense of are the relatively primitive taxa (i.e., in relation to obviously derived taxa); sampled organisms in such plesiomorphic groups do not carry specialized attributes that allow their ready differentiation. Reptiles are a case in point: all the features that appear to unite living tetrapods into a group called *Reptilia* are primitive with respect especially to birds; not surprisingly, some reptiles (e.g., crocodiles) are more closely related to birds than are others (e.g., turtles).

On the other hand, patently monophyletic taxa are detected by the joint possession of phylogenetic markers, or organismic attributes. Birds share, among other features, feathers, which are a derived condition of the more general tetrapod vertebrate feature, epidermal scales. Like well-marked clones, monophyletic taxa are nothing more—nor less—than phylogenetically linked series of species sharing one or more organismic features that serve to mark a particular lineage. There is nothing more to the origin of monophyletic taxa, then, than the first appearance within a species (whether or not at the origin of that species) of one or more organismic attributes that are subsequently passed along to further descendants for the duration of the lineage. Speciation is the highest level of "more-making" that can be recognized in the biological realm.

Because higher taxa, themselves formed as an outgrowth of reproductive activities, cannot be said to engage in reproductive activities in any meaningful sense, we have reached the highest rung in

TABLE 4.1.

The Genealogical Hierarchy

Monophyletic taxa
Species
Demes
Organisms
Germ line

a system of categories of entities that compose the genealogical hierarchy (table 4.1). Each category represents a set of entities whose parts are unified by the reproductive (i.e., "more-making") activities of their component, constituent parts: organisms reproduce, keeping local reproductive communities (demes) going. Demes come and go, merging and splitting, and providing the internal structure of species, as Wright so convincingly saw. Species are kept going by the production of new demes (or, in the absence of evident demal structure, as when a species is reduced to a single local deme, simply by the ongoing reproductive activities of component organisms; for each category, it is possible that an entity will have but a single component part, rather than many—except species, which are defined here as inherently and exclusively a reflection of sexual reproduction, thus requiring minimally one male and one female). Speciation, finally, produces and maintains monophyletic taxa. Monophyletic taxa do not make more monophyletic taxa; the monophyletic taxon *All Life* is at the apex of the Linnaean hierarchy. And thus there are no further categories of the functional genealogical hierarchy beyond monophyletic taxon. For each level of the genealogical hierarchy, component entities are maintained (cohered) by the reproductive activities of their constituent parts, and their own reproductive activities make them parts, lending coherence, of entities at the next higher level.

We have yet to consider the somewhat ambiguous status of demes. After all, local populations can be construed in both an ecological and a reproductive sense. It is obviously so that for any given species, its local populations may be simultaneously demes and ecological (economic) entities; we shall follow Damuth and call the latter avatars. In other words, there may be something approaching formal identity, at least in some situations, between demes and avatars—local populations in the genealogical (reproductive) and ecological (economic) senses, respectively. In other words, the burgeoning split where organisms remain economic and reproductive kinds of entities, no matter how great a division of internal labor may obtain within any particular kind of organism, may not be as complete at the local population level as it manifestly is at the species level (purely genealogical) or at the ecosystem level—to cite a patently nongenealogical (i.e., distinctly cross-genealogical), economic kind of biotic system. It is clear that local systems and their potential blend of economic and reproductive activities deserve particularly close attention—the more so because social systems are highly peculiar blends arising from amalgamations (of various kinds) of organismic reproductive and economic activities. Before we tackle the demes/avatars situation in depth, however, we first need to take a more general look at the biotic consequences of organismic economic activity.

• • • • • • • • •
Organismic Economics and Biotic Organization

Sexual reproduction, of course, requires more than one organism—usually two, though in some mating systems (e.g., king snakes) more than one male may be involved with the insemination of the eggs of a single female. More to the point, however many partners are involved in sexual reproduction, is that the very existence of the system implies a pool of prospective mating partners. Sex begets biotic structure and requires its presence for maintenance of a given volume of successful matings to continue.

Not so, at least at first glance, with economic[1] activities. Most organisms develop, grow, and maintain physiological homeostasis as "stand-alone" matter-energy transfer systems. They are independent—not of other organisms in general, but of a need to be directly associated with other organisms of immediate like kind when it comes to the most fundamental physiological aspects of maintaining a somatic physiological plant. It is possible, in other words, to eke out an existence as a hermit—but not, in so doing, to leave any children behind.

Yet even photosynthesizing plants, despite their apparent economic self-sufficiency, generally require the presence of other organisms for their own survival. And certainly, in the realm of animal heterotrophy, organisms rely utterly, in an economic sense, on the presence of other organisms. Economic behavior, then, does indeed imply biotic structure, and the most immediate aspect of such structure is that it is inherently and characteristically cross-genealogical. Indeed, it is the interaction of many different kinds of economic activity that yields the energy flow that is the very life blood of ecosystems.

Yet before we consider in greater detail the cross-genealogical structure of local and regional ecosystems, we must consider the possibility that—the potential for economic autonomy *within species* aside—there may be something about characteristic economic behavior in general that fosters conspecific, local biotic structure. We have argued at length in the preceding chapter that large-scale entities arising as a consequence of reproductive behavior—species—are not themselves to be construed as economic entities. We also argued that shared similarity in economic adaptations within reproductively defined species does not suffice as a dynamic contributing to the actual cohesion of those entities we call species. We now consider whether local populations of conspecifics—demes, when seen as arising from reproductive behavior—have an economic analogue, as is traditionally recognized in ecological theory.

Heterotrophic economic behavior clearly implies a form of interaction among local conspecifics. The most obvious point is that conspecific heterotrophs, as the overwhelming rule, do not prey on one another. Local conspecifics are not legitimate sources of energy; indeed, many of the exceptions to this generalization involve parental consumption of young rather than systematic destruction and consumption of local nonkin—something of an anomaly when viewed in ultra-Darwinian terms. Yet the generalization that conspecifics do not ordinarily constitute food items makes perfect Darwinian sense: no system could be stable, or remain extant for long, if the seeds of its own destruction were firmly implanted in its behavioral ecology. Among carnivorous heterotrophs, avoidance of conspecifics as prey ranks as a definite aspect of ecological interaction.

More general, and canonical to both Darwin's and Wallace's formulations of natural selection, is the notion of competition. Carnivores may not kill each other for direct consumption, but all organisms may be expected to compete with local conspecifics for energy resources. Such competition arises as a simple consequence of local conspecifics sharing a great deal of genetic information, most of which pertains to economic attributes. Local conspecifics, in other words, all do pretty much the same thing for a living—and will be more similar to each other in that respect than they will in general be to other nonconspecific organisms. Trees of the same species, if rooted adjacent to one another, will compete for sunlight and for nutrients from the soil, because they grow at similar rates, reach similar characteristic heights, have similar capacities for developing characteristic root systems, and have closely similar physiological needs. Competition within local populations of conspecific trees is every bit as real a phenomenon as is the more obvious competition between conspecific heterotrophs for food items.

Traditional, too, in the Darwinian heritage is to see energy resources as perhaps the most important limiting factor of local conspecific population size. It is conventionally assumed that local populations are at or near their maximum population size, given the "carrying capacity" of the local environment, that is, the energy resources available to organisms with a given set of economic adaptations.[2] Such a vision of the factors governing population size necessarily entails a high degree of within-species competition: populations at or near maximum size are presumed to be stretching resources to the limit, virtually forcing organisms to compete. In this, the prevailing model of local-population economic dynamics, competition emerges as the strongest component of conspecific organismic interaction.

Yet it is clear that other aspects of conspecific economic interaction exist. There are patterns of avoidance and subdivision of territories—usually discussed in conjunction with reproduction, but a matter of economics as well. Male cats in all species except cheetahs and lions habitually live alone, marking the boundaries of the extensive territories. The effect of this territorial behavior, at least for the greater part of a male cat's existence, is to provide hunting grounds with a minimum of (direct) competition with other males.

Beyond neutrality and the economically positive effects of resource subdivision through territoriality lies the realm of cooperation. Economic cooperation, of course, is a hallmark of social systems generally—though in seeing the origin of eukaryotes through multiple events of phylogenetic symbiosis, Margulis and Sagan (1986) argue that cooperation is a more dominant, pervasive theme in nature than is competition and is not restricted to social systems. The basic thesis of this book is that social systems are peculiar, idiosyncratic biotic outcomes of the reintegration of organismic economic and reproductive adapatations. We see such reintegration arising from a more general background of a functional distinction between economic and reproductive activities in the daily lives of nonsocial organisms. Most examples of conspecific economic cooperation that spring readily to mind (such as pack hunting behavior of African hunting dogs) involve frankly social organisms. In such cases it is difficult to disentangle the economic from the reproductive in social systems; indeed, that is the whole point—social systems represent labyrinthine complex relations between economic and reproductive behavior.

As we have already seen, sociobiology usually interprets all economic activity of social organisms as mechanisms to maximize the genetic fitness—the reproductive success—of individual organisms. Economics, in this mode, is *really* all about reproduction. We need to ask a simpler question: is there any evidence of economic cooperation within local populations of conspecific organisms that can fairly be viewed as nonsocial?

There is reason to suggest that the mere presence of local conspecifics may enhance a given organism's economic life. Presence of past and contemporary conspecifics often prepares the environment—as when well-established coral reefs offer footholds for new colonies to become established. To be sure, there are many instances in which larger colonies established at an earlier time will crowd out younger colonies (as Winston 1990 describes for intertidal bryozoan colonies); yet it is the prior success of conspecific (and nonconspecific) organisms that provides the reef environment in the first place, enabling larval recruiters to find suitable habitat to become established and grow—to lead their economic lives.[3] Similarly, another example of conspecifics literally paving the way for later organisms is the development of systems of trails (such as those used by deer on Komodo Island in Indonesia; Komodo dragons lurk along these trails, lying in ambush for prey items, according to Auffenberg (1981) between feeding grounds, water sources, and so on.

Organisms, as Lewontin (1983) has been particularly concerned to argue, really do, in part, create their own environments—and such modifications have direct implications not only for the particular organism creating the modified environment, but also for other organisms, conspecific as well as nonconspecific (as when burrowing owls occupy prairie dog holes—though that is primarily for nesting!). It is probably straining the language to characterize such positive economic effects among nonsocial conspecifics as cooperation. Nonetheless, there do seem to be reasons why nonsocial conspecifics live in close proximity beyond the fact that they are attracted to the same environments on economic grounds and their occasional need to reproduce.

The SMRS, almost ironically, is relevant at this juncture in our consideration of conspecific organismic economic interaction. When swallows rest on telephone lines between feeding forays, they tend to cluster together. When more than one species is represented, the birds tend to segregate as the old adage would have it: barn, bank, and tree swallows form subclusters within a large mixed congregation of resting swallows. There is every reason to assume that the criteria of recognition of conspecifics are the very same in an economic situation such as this as in frankly reproductive circumstances. Among complex animals, the visual cues of actual recognition of mates (and of sexual competitors) also lead to recognition in economic contexts. That the SMRS is the more general property holds because *all* sexually reproducing organisms possess a species-specific fertilization system. Among truly motile complex animals, where an element of actual perceptual recognition (i.e., over and above various mechanical and chemical elements of recognition in mating systems generally) is added, the system is readily co-opted to allow recognition of conspecifics in a purely economic context. Such an "SRS" (i.e., simply "species recognition system") is a manifest prerequisite to the formation of even the most rudimentary social system.

Economics, Populations, and Niches

The degree to which local populations of conspecifics are naturally unified by purely economic interactions is one issue; the degree to which local populations of conspecifics act as economic entities, while related, is another question. Even if local conspecifics are there simply because they share a preference for a particular habitat configuration, it is still possible that the population as a whole exerts a unitary economic effect on the abiotic and biotic components of the environment. Such is, in fact, the traditional view of ecosystems ecology: it is local populations of conspecifics that have "niches." Whether the niche of a local population arises as a simple summation of the economic behaviors of the component organisms, or whether those organisms are interacting economically to form some larger, coherent entity, it is the characteristic, repeated patterns of economic behavior of organisms of various different species that determines the interactive nature of local ecosystems. Density (population size) is especially critical here. Energy flow patterns change in complex, mutually interdependent ways as population sizes oscillate—as in the well-known examples of predator-prey cycles, in which population explosions in prey lead to increase of predator population, until prey population declines enough to lead to a crash in predator population size.

Though this discussion has leaped to the realm of cross-genealogical interaction, the point remains that from a systems point of view, it makes good economic sense to think of unitary roles of local populations of conspecifics. A red fox eats a ruffed grouse, with varying implications for the economic and reproductive success of both organisms. But it is the economic role (only in part) of red foxes to dine on ruffed grouse, and for the latter to be wary of foxes as an economic hazard; the frequency of such events depends upon density of populations of each kind, which is in turn dependent on many additional factors.

The economic concept of *niche* refers to the characteristic role played by a particular kind of organism in a local ecosystem. Yet organisms do not have niches; organisms of like kind within functioning ecosystems are said to have niches. It is, really, local populations of conspecifics that have niches in local ecosystems. Precisely because we can ascribe an economic role to local populations of conspecifics within ecosystems, it is useful to adopt Damuth's (1985) term *avatar*. Demes are local populations in a reproductive sense; avatars are local populations in an economic sense. We have already briefly seen how avatars and demes may, but definitely need not, be the same in any particular instance; we return to the relation between these two sorts of local aggregates of conspecifics when we turn to social systems per se. We now take a brief, closer look at the integration of avatars into functional local ecosystems.

Avatars and Local Ecosystems: Up the Economic Hierarchy

When we confront the structure of economic systems formed through the interactions among avatars and between organisms and avatars and the physical environment, the distinction between economic and reproductive biotic systems becomes starkly realized. It may, in any particular instance, be difficult to distinguish between an avatar and a deme. Each is, after all, composed of conspecific organisms.

As we have already noted, the same organisms often are involved in each, so the distinction between an economic and a reproductive functional entity at the local population level may well be conceptual rather than a reflection of a real difference between two distinct entities. Social systems, of course, simply reflect further erosion of any meaningful distinction between avatar and deme.

Such is not the case when entities at the next higher level within the economic and genealogical hierarchies are compared. Demes are parts of species; avatars are parts of ecosystems (but see below). Species can not be confused with ecosystems; nor are species *parts* of ecosystems—though one frequently reads lists of species present in given ecosystems, or simply in geographic regions or habitats. In such instances, ontological confusion creeps in as a consequence of accustomed linguistic usage: when we speak of "the number of species visiting a backyard bird feeder in the wintertime," we mean that one or more organisms of each "number of species" were seen visiting the feeder. *Species* do not visit feeders, *birds* do—and each bird belongs to a local population of a given species. Note, too, that even though the birds (especially if they are year-round residents) may represent a sampling of a local deme, the context of winter feeding is decidedly economic—yet we enumerate the birds with their species names. Taxa, the higher ranges of the genealogical hierarchy, are simply accumulations of phylogenetic history, and hence repositories of genetic information—thus they are the logical units for the enumeration of organisms even when the organisms are being considered (named and analyzed) in a purely economic context, as when the different species present in a local ecosystem are being enumerated.

Whatever the entanglement between genealogically based nomenclature of elements of economic systems might be, it is clear that large-scale biotic systems arise purely as an effect of the economic activities of organisms. Hierarchy theorists are divided on the issue of the correct basic approach to enumerating cross-genealogical, upper-level elements of the economic or ecological hierarchy. The primary issue is one of purity of biological entities. Thus some (e.g., Salthe 1985; Pilette 1989) prefer to speak of *communities, regional biotas,* and similar entities, because such units contain only biological systems as parts; for example, local populations (in an economic system) are parts of communities. Others (e.g., Eldredge 1985) prefer to use terms such as *ecosystem* (e.g., "local ecosystem," and the more inclusive "regional ecosystem"); such concepts explic-

TABLE 5.1.
The Ecological Hierarchy

Biosphere .
Ecosystems
Avatars
Organisms
Soma

itly include the abiotic environment along with purely biological components. The argument for such usage is that the economic *work* of organisms, and how that work is translated into the formation of larger-scale economic systems, is embodied in the concept of ecosystem. The notion of energy flow, which is seen to connect the variegated elements of any well-analyzed ecosystem in an intricate web, lies at the heart of this concept. It is this very form of interaction that coheres economic biotic systems—a form of interaction that differs drastically and dramatically from sexual reproduction and the kinds of biotic interactions that sexual behavior entails.

Thus a simple consequence of interaction—*economic* interaction involving energy flow—is that complex, coherent biotic systems are formed. The sorts of interaction,[4] moreover, that unify ecosystems is different in kind from those that unify avatars. The economic needs of organisms within avatars coincide, leading to competitive or cooperative forms of direct interaction (as well, of course, to a neutral mutual tolerance, itself a subtle form of interaction). It is another matter entirely between avatars. There are autotrophs and heterotrophs (as well as saprophytic organisms), herbivores and carnivores. With the minor exceptions alluded to briefly above, energy flow is *predominantly between*, rather than *within*, avatars.

A further word is in order on how energy flow between avatars unifies local ecosystems and unites local ecosystems into regional economic systems. Recent discussions of the nature, composition, and relations between the economic and genealogical hierarchies often visualize ecosystems as arising purely from the interactions among the next lower component parts—in this case, among avatars. This is the position, for example, adopted in Eldredge and Salthe (1984), as well as in Salthe (1985) and Eldredge (1985). Yet it is individual organisms that are engaged in economic processes—

most certainly those interactive (and cross-genealogical, for all but autotrophs) processes of energy procurement. In other words, a case could be made that certain types of nonreproductive interaction (between local conspecifics) unifies local economic populations (avatars), but that a different set of (cross-genealogical) interactions unifies ecosystems. Such an argument conflicts with the view that functional hierarchies consist of elements unified at any given level through the interactions of their next lower constituent parts—and so on on down the list of levels. For such a view to be maintained in the economic hierarchy, the soma of an organism would be seen to be maintained by interaction and (literal) cohesion among its parts; avatars united by local conspecific organismic interaction; and local ecosystems cohered by interavatar interaction. This last view is obviously less reductive than the view that sees ecosystems as a simple reflection of (nonspecific) interorganismic interaction.

Do avatars, then, really interact (as we have already maintained they do earlier in this chapter)? Of course, individual foxes prey on individual ruffed grouse. But it becomes meaningful to see avatars themselves as actual interacting entities when we see the intensity and rate of interaction as regulated by distributional densities of organisms within avatars. And distributions are properties of the larger-scale entity; just as spatiotemporal distribution of species is a species-level property, the rate of economic interaction among nonconspecifics in a local ecosystem is governed to a large degree by the spatiotemporal distribution characteristics of the local economic system. In this sense, the dynamics of energy flow within local ecosystems is governed by interavatar, rather than simply interorganismic (nonconspecific), interaction. Thus it can be maintained that niches—economic roles—are best construed at the avatar, rather than simply at the single organism, level.

Cohesion and the "Reality" of Ecosystems

Biologists are by no means in universal agreement on the ontological status of economic associations of organisms. The debate centers around several distinct focal issues: the extent to which biotic entities are truly unified through economic interactions, the permanence of such entities, and the discreteness that such entities may or may not display. "Real" entities (individuals in the sense of Ghiselin 1974a and Hull 1976; see Eldredge 1985), after all, should be internally coherent and spatiotemporally bounded. The debate about the reality of ecosystems mirrors exactly that about the reality of species, in that the underlying issues are the same: are (species, ecosystems) merely collections of organisms that share some properties (anatomical similarity for species; sympatry for ecosystems)? Or are they entities, systems bound together by virtue of some activities (interactions among components), and bounded both spatially (i.e., geographically) and temporally—in short, entities with beginnings, histories, and terminations, with discrete distributions in space at any given moment of their existence?

Ecosystems are invariably discussed in terms of networks of energy transfer: energy flow provides the cohesion one would demand of a real biotic entity. But often the reality of cross-genealogical systems is discussed in a slightly different context: communities, rather than ecosystems (with their additional abiotic components), are the focus of attention. Communities themselves are variably construed; often what is meant by *community* is simply "all the birds living in a specified, contiguous area." But the term also applies to *all* the biota living in a particular, specifiable habitat. In particular, communities are commonly seen as recurrent assemblages of (parts of) a particular collection of species. In other words, similar cross-genealogical associations of organisms tend to crop up within a region whenever appropriate conditions occur. Two species of three-toed woodpecker, boreal chickadees, Canada jays, kinglets, spruce grouse, and a number of other species occur in spruce forests well up the slopes of the higher peaks of the Adirondack Mountains of New York State; similar associations also occur at lower elevations around peat bogs (at the margins of lakes with high natural acidity) with tamarack and spruce.

Some ecologists see such recurrent assemblages as a mere reflection of subsets of locally occurring organisms happening to share similar habitat preferences. Zoologists tend often to view plants as part of the background habitat; if attention is given only to some portion of a recurrent assemblage, it may well be natural not to expect to find a high degree of accommodation (economic interaction, especially, but not limited to, energy flow) when all biologic components of recurrent assemblages are considered at once. Viewed in this way, natural associations of organisms, whether seen explicitly as ecosystems or simply as communities, are routinely acknowledged to display at least some degree of biological accommodation—a cohesion supplied by cross-genealogical, economic interactions among component parts. The debate then usually centers on the degree to which communities so construed display such accommodation; the consensus appears to be that some communities are more tightly bound together by such interaction than are others.

Indeed, it was on the basis of positing such a spectrum of accommodation that Stenseth and Maynard Smith (1984) contrasted the two fundamental evolutionary patterns they claim operate in ecosystems: one (the "Red Queen"; cf. Van Valen 1973) sees species (meaning, of course, local populations of species) constantly changing in response to changes in other species. In contrast, their "Stationary" model envisions evolutionary change occurring only in response to physical (abiotic) environmental change—with stability the rule in the absence of concerted environmental change. The Red Queen reflects highly accommodated ecosystems, while the stationary model is claimed to be typical of ecosystems where levels of biotic accommodation (interaction) are quite low.[5]

It seems to us that there is a spectrum of process leading to the development of recurrent assemblages of organisms—a spectrum ranging from mere accidental concatenation born of shared environmental preferences, on through assemblages whose component populations are so tightly interdependent that each is necessary for the assemblage to persist. At the same time, it must be true (indeed, almost by definition it *is* true) that energy flows among nonconspecific organisms in all local habitats where more than one (or at most a very few) species are represented. That being the case, it is safe to postulate that all communities, or ecosystems, possess some degree of accommodation, that is, the system exists in part because its specifiable component parts display a very real, dynamic form of interaction.

Beyond Cohesion: Spatiotemporal Bounds of Economic Systems

An entity (an individual in the sense of Ghiselin 1974a) need not require dynamic interaction among its parts to be seen to exist, to be real. A pile of bricks sitting in a yard is held together by gravity and by friction between parts. The shape and surface texture of the individual component bricks determine the angle of repose of the pile's outer surface. But there is nothing further intrinsic to the individual bricks that acts interactively to bind the pile together. And while few except a building contractor would care to consider the reality of a pile of bricks, it seems clear that it is the discreteness, rather than any consideration of internal cohesion, that forms the strongest basis for claiming that a pile of bricks constitutes an entity. The issue is important precisely because the argument most commonly developed against seeing biotic economic systems as real entities hinges on the lack of discrete boundaries between adjacent ecosystems. For every example of the sharp boundary between pond margin and surrounding subaerial habitat, there are countless situations of large tracts with subtle variations in physical conditions and resultant biotic distribution that defy easy characterization into discrete ecosystems. Purely at the biotic (community) level, some species are represented in a greater range of habitat types (hence in more communities or ecosystems) than others. Coyotes range from coast to coast, from desert to forested mountains; probably no single species among their varied prey (primarily rodents, rabbits, and birds) ranges as far afield.

This lack of tight discreteness, with few definitive boundaries between ecosystems whose components are rigidly restricted within them, has fostered the belief that economic systems are classificatory figments of the ecological imagination. Allen and Starr (1982), moreover, believe that the real boundaries between entities composing a system may well be other than those we posit; they persist in seeing economic biotic nature nested into a hierarchical array of such systems, however, for the heuristic benefit that such a scheme lends to approaching the otherwise bewildering complexity of natural economic systems.

Allen and Starr's example to illustrate their point that conventionally attributed boundaries to economic sytems may not correspond to the actual organization of economic nature centers on organisms: though organisms (generally, excepting oddities such as the slime molds) appear to have discrete boundaries, such as those afforded by the skin of vertebrate organisms, there is a constant interchange of gases, liquids, and solid materials (as when dead skin cells are regularly sloughed off) with the surrounding medium, dynamically blurring the picture of discrete boundedness. Allen and Starr (1982) conclude that the actual outer bounds of an organism might lie some inches off the apparent surface of the skin.

Thus the issue of boundaries of systems appears closely linked to the epistemological issue of scale of observation. We must remember that ecosystems, like species, are large-scale entities, entities of which organisms are parts. We are organisms. Although the remote sensing of ever-smaller bits of the material universe is limited only by the current state of technology, and apart from the true remote sensing of satellites affording a bird's-eye view (and beyond) of physical and biotic habitats, there is little comparable technological equipment to which we can turn to help us see systems with such large-scale spatiotemporal characteristics. Boundaries between ecosystems are inevitably going to appear fuzzy to us. There is an analogy here with the problem of seeing cohesion: we persist in seeing atoms as fundamental building blocks of matter, despite the fact that it has long been established that atoms are mostly empty space. At the very least, there is no difficulty in seeing atoms as real entities—even, albeit somewhat paradoxically, as solid. But we see the component parts of species, and of ecosystems, as separate entities, and it is by no means apparent to all that the interactions among these parts result in the cohesion of entities that are every bit as real as are atoms.

So, too, with boundaries. Where is the outer bound of an atom? There is no neatly defined outer shell—just a blur of chaotically occurring electrons of various energy states; and the overall shape of this blur is further subjected to extreme modification as the atom interacts with others of the same or different kind in chemical reactions. Yet if atoms aren't real entities, where are we in our description of the contents of the material universe?

Thus we are firmly aligned with that substantial set of ecologists who are not especially troubled by the readily acknowledged difficulties inherent in detecting boundaries of ecosystems. Ecosystems indeed leak—organisms are constantly ranging from one system to another. Such movement blurs the edges of the systems involved, but it also constitutes the prima facie evidence that local economic systems are themselves interactive: matter and energy flow between adjacent systems in a network that is simply a higher-order version of the same sorts of matter-energy transfer patterns routinely diagrammed by ecologists (for examples and discussion see, e.g., Ulanowicz 1986) within ecosystems.

Thus there is a strong and fundamental geographic component of economic systems. Species, too, have finite (if, at times, discontinuous) distributions. Different species may have virtually the same distributions, though the ranges of any two species are seldom if ever precisely identical. Not so with economic systems: there can be but a single economic system at any one time and place on the face of the earth. One ecosystem may entirely surround another (as when a forest encircles a lake), but it is the nature of economic systems, quite unlike genealogical systems, that only one such system can be present at any one place and time.

There is one other aspect of distribution that seems to be an essential ingredient of the reality of any given putative entity: not only are entities spatially bounded, but they are also temporally bounded. Organisms are formed (for some, conceived, developed, then born, hatched, or sprouted), live for some time, and eventually die. Species arise, persist, and become extinct. All entities have histories—even elementary particles, whose evanescent histories are measured in nanoseconds. Such is true, as well, of economic systems: habitats supporting more or less the same aggregation of component organisms (species representatives) persist from year to year during our lifetimes. It remains, however, for us to explore the significance of the historical nature of ecosystems, especially in relation to the historicity of genealogical systems.

The fact that economic systems change through time has been cited (along with difficulties in precise boundary determination) as evidence that such economic systems do not constitute real entities. The equivalent argument is, in fact, the canonical line of thought (e.g., Bock 1979, 1986; see Eldredge 1985 for further discussion) underlying the argument that species, viewed through time (i.e., as historical entities) also are not real. Through time, the anatomical properties of organisms within species are necessarily and inevitably transformed: species evolve themselves out of existence in this manner—and can hardly be said to be real entities. Ghiselin (1974a) and Hull (1976) turned the argument on its head: according to them, because there is an inevitable transformation of organismic (genetically based) phenotypic properties through time, species *must* be construed as real entities—as "individuals," as they put it. Because the properties of species are constantly undergoing evolutionary modification, species can in no sense be construed as a "class" with certain immutable defining properties. They must be something else—individuals—though, as we have noted, there are some real difficulties with this choice of terminology.

Nonetheless, the fundamental distinction drawn by Ghiselin and Hull is important, valid, and relevant for a consideration of many sorts of biological systems other than species. Organisms, those prototypical individuals, are constantly changing between birth and death. Ecosystems go through characteristic stages of succession. Succession is arguably analogous with ontogeny of organisms, though detailed comparisons of both processes would, no doubt, reveal many differences—such as the fact that ontogeny represents an interplay of genetic and environmental information, and genetic information is far more directly and heavily implicated in ontogeny than could ever be maintained for ecological succession. The very fact that ecological systems are cross-genealogical precludes the systematic connection between ecological and genetic systems: there can be no set of genes for ecosystem succession—a point to which we will return.

But ecosystems change through time in ways having little or no relation to the more or less orderly processes of succession. Organisms come and go; some species of birds show up as casual vagrants from Eurasia on both coasts of North America every year. The local pair of hairy woodpeckers might disappear over a winter, and hairy woodpeckers might not be seen in a particular woodlot for several ensuing years, if ever again (though given the function of species as suppliers of players to the economic game, most such disappearances are only temporary; see Eldredge 1986).

It is especially this local change in species representation—the disappearances, or invasions of newly arrived representatives of additional species—that have led some biologists to add such fluctuations in composition to boundary fuzziness when they conclude that ecosystems are arbitrarily defined heuristic constructs and are not to be taken seriously as real entities. Yet the component parts of systems can change, and the entity, the system, itself is perceived to continue. This precise argument was broached by Caplan and Bock (1988) to show that biological systems cannot be correctly construed as individuals: citing the New York (football) Giants, and pointing to the ever-changing roster of players on the team, they note that the Giants nonetheless persist, have had a history, continue to play games, and will do so into the future. Caplan and Bock view a football team as a class; Ghiselin, more to the point, would see them as a corporate entity. To Ghiselin, General Motors and the New York Giants are patently individuals.

We therefore expect there to be turnover in species representation within ecosystems—and we see that phenomenon as no threat to the conceptualization of biotic economic systems as real entities. Simberloff (e.g., 1972) has investigated many of the processes that alter the composition of local economic systems; migration rates are clearly important, as are the tolerance limits for both the addition and deletion of components. Depending upon degree of biotic accommodation within any given system, it is notorious that loss of some species may well entail serious consequences for the survival of the entire system; in other instances, loss (whether or not it is temporary) will have no palpable effect on the system. Similarly, presumably because of competition, many invasions last only temporarily—permanent invasion is often difficult or impossible. Yet many of the recently documented range expansions in North America (e.g., northern migration of many species, such as opossum, cardinal, and tufted titmouse) appear to be genuine invasion of new habitat rather than expansion back into formerly occupied territories.

Thus, checkered as it may be, the history of any ecosystem hardly debars its consideration as a real system. But there remain further issues of historicity in economic systems: specifically, why do biologists seem to care far more about the histories of genealogical systems (taxa: species however construed, and monophyletic clusters of species) than they do about the histories of local and regional economic systems?[6] The answers to this general question reflect the actual differences in the natures of these two different kinds of entities.

Further Aspects of Historicity in Economic and Genealogical Systems

Damuth (1985), as we have already seen, established the term *avatar* for local representation of a species. Although Damuth's discussion clearly links avatars with economic activities, nonetheless the thrust of Damuth's paper is the analysis of higher-level selection: avatars are both interactors and "more-makers," and it is they, not other biotic entities (especially not species), that are the focus of higher level selection.

Damuth (1985 and pers. comm., 1985) argues that new avatars arise from old: though avatars form parts of local economic systems, they also fulfill Hull's second criterion of selection (replication—extended simply to more-making, e.g., in Eldredge 1985). We prefer to distinguish demes from avatars, thus continuing to see distinct aspects of economic interaction and reproductive processes at the local population level—while readily conceding that in instances where avatars and demes closely approximate identity, selection at that level becomes readily plausible. As we discuss in a later chapter, because organismic economic and reproductive activities are intricately associated in colonies and social systems, biotic selection at these levels emerges as a virtual certainty.

We briefly raise issues of selection in economic systems here because, as in true genealogical systems, there is a nested set of resemblances linking economic systems in both space and time. The truly genealogical nested set of taxa of the Linnaean hierarchy has universally been conceded to be *the* fruit of the evolutionary process ever since Darwin (1859) pointed out that a nested set of anatomical resemblances would be the naturally expected outcome of a process of "descent with modification." The very notion of evolution explains the existence of the hierarchical organization of life—a pattern that had been recognized and accepted long before phylogenesis became its accepted causal explanation.

But, in a distinctly economic sense, there is also a nested pattern of resemblance linking biotic associations. Nineteenth-century biologists spoke of "geminate" species, or of "vicars"—alluding to allopatric distributions of (generally) closely related species appearing to play closely similar, analogous roles in disjunct ecological settings. The fossil record provides even more striking examples: similar ecological assemblages, many of whose component taxa are close relatives, are often found to replace one another geographically. Imbrie's (1959) analysis of the brachiopod fauna of the Devonian Traverse Group of the Michigan Basin reveals many assemblages that are closely similar to others found in the Hamilton Group and equivalents elsewhere in eastern and central North America. The species in the Michigan Basin are closely allied phylogenetically with others elsewhere (when, indeed, they differ at all).

Similarly—and this is where the point begins to have real force—analogous (ecologically similar) faunas occupying comparable habitats tend to replace one another through time. Over tens of millions of years, there can emerge a hierarchical array of ecological similarity linking a long succession of faunas. For example, near-shore bottom environments of the mid-Paleozoic era display characteristic recurrent suites of mollusks and other marine invertebrates (Bretsky 1969). Comparison of particular subsets—for example, those occupying sandy, high-energy environments—reveals assemblages that remain remarkably similar for periods of one hundred million years or more. Those near-shore sandy environments, from the mid-Ordovician on through the mid-Devonian Period, time and again produce an association of large, scalloplike bivalves, wide-apertured bellerophontid gastropods, rhynchonellid brachiopods, and large homalonotid trilobites. Homalonotids became extinct after the mid-Devonian Period, but analogous assemblages of many of the other components persist into the Carboniferous Period.

The resemblances between such economic entities are clearly phylogenetically based. (We are not, in other words, talking simply of analogous economic roles played by phylogenetically dissociated members of different ecological systems.) The further removed in time and space two analogous ecological settings are, the more remotely related phylogenetically their components will be. Thus the nested pattern of resemblance linking ecological systems is actually phylogenetic relationship: the homalonotid trilobites of the near-shore faunas in the Devonian Period (different places, and different times in the Devonian Period) are generally more closely related to one another than they are to the Ordovician homalonotids (though such need not be the case, as when sublineages of a clade persist through long periods of time).

It was fashionable in the 1960s to speak of "community evolution." The nested pattern of similarity linking ecological systems in space and time formed the basis of such thinking. As we have seen, this pattern is actually phylogenetic *at the level of individual avatar components:* extant members of clades tend to form ecological assemblages similar to those formed by their phylogenetic progenitors in earlier times. But is this truly "evolution"? Or is it simply the accumulation of history—a long series of persistent organization using progressively less and less similar ingredients to form functional, economic assemblages? *We have difficulty imagining the "evolution" of any economic system because there seems to be no systematic way that cross-genealogical information can impinge on genetic information.* Put another way, there is no process of information storage in economic systems that is comparable to that in phylogenetic systems. It is our view that the nested pattern of resemblance linking ecological systems into a historical ecological hierarchy (the "Bretskyan hierarchy"; Eldredge 1985) results purely from the moment-by-moment functional interactions among cross-genealogical elements that repeatedly assemble in comparable environments. The similarity simply reflects the phylogenetic pedigree of the component avatars. Damuth (1985) might be right in asserting that avatars display both ingredients necessary for selection to occur, but cross-genealogical, economic systems cannot impinge systematically on genetic information—and thus, or so it seems to us, there can be no process of ecological evolution in any meaningful sense of the term.

There are still further issues of historicity versus functionalism in modern science that we shall touch on briefly, if only because they have provoked much misguided rhetoric. Some authors have even claimed a fundamental difference between various branches of science because some deal with the particulars of history, while others focus on classes of events and processes. We merely note here that the valid distinction between histories of particular systems and the study of classes of events and processes does not exhaust the issues—nor does it truly divide one branch of science from any other. All entities have histories; chemists have yet to find a particular use for studying the history of any single water molecule, but such could be done in principle (for newly formed molecules, at any rate). Closer to home (in evolutionary biology), population geneticists studying generation-by-generation data pertaining to gene frequency changes are indeed studying historical data. But they are doing so in any instance as an exemplar that helps to elucidate general processes governing rates and other aspects of genetic change within populations. They do so, in other words, in the spirit of investigating classes of events and process. As Ghiselin (1987:129) has recently pointed out, there are generalizations ("laws," as Ghiselin call them) of this nature at all levels of evolutionary biological discourse, despite Mayr's disclaimer. Ghiselin's example is telling— "Allopatric speciation is such a law"—and, as Ghiselin says, no one has discussed allopatric speciation more extensively than Ernst Mayr! Species, like water molecules, keep coming into existence.

Distinctions—and Interactions—Between Economic and Genealogical Systems

We have thus far followed the implications of two separable aspects of organismic interactive behavior for the formation and maintenance of two different sets of biological systems. We have noted, though not belabored, that each of these two sets of systems is hierarchically nested: sexual reproduction leads to the establishment of a nested set of reproductive communities that are, at base, progressively larger packages of genetic information. Ongoing economic interaction leads to economic associations of local conspecifics, themselves parts of local, then progressively more regional, economic systems.

We have also noted that the dichotomy between these two systems is realized in nature (i.e., is not simply an artifact of human thought, whatever its value as a heuristic may be). But the schism between economic and reproductive systems is not absolute, nor does it pertain with equal force at all hierarchical levels of biological organization. It is clear, for example, that organisms are reproducers *and* economic entities. Reproduction costs energy: there are natural connections between economics and reproduction.[7]

We have already discussed in detail the implications for understanding selection processes of recognizing the dual status of organisms as economic and genealogical entities. In particular, natural selection is the natural link between the two hierarchical systems. There is a vector from the economic world to the reproductive for organisms within avatars/demes: relative economic success has a statistical effect of relative reproductive success. Furthermore (and as discussed in greater detail in Eldredge 1985, 1989), there is a feedback vector: genealogical systems, especially species organized as a series of quasi-isolated semiredundant demes, provide the supply of players to local ecosystems—a phenomenon most graphically seen after ecosystem degradation, ranging from the local effects of (relatively minor) fires, oil spills, and so on, to recovery from the global effects of mass extinctions.

We have further recognized that simpler organisms present far less of a formal, physical split between economic and reproductive structure and function than is apparent in more complex organisms—organisms, moreover, belonging to taxa that appeared later in phylogenetic history. Indeed, a burgeoning split between the economic and genealogical realms grows with phylogenetic history.

Above the organismic level, demes and avatars may be composed of quite different sets of organisms depending upon the time of year. In other instances, it may make more sense to speak simply of a local population whose composition does not vary much, if at all, regardless of time of year, and specifically whether or not reproduction is taking place in addition to ongoing ecological interactions.

Only when cross-genealogical ecosystems are contrasted with higher-level genealogical systems—species and higher taxa—does the dichotomy become vivid and nearly absolute. Mere tendencies towards compartmentalization at lower levels result in two completely separate systems by the time these higher levels are reached. Organismic economic interaction drives one system, reproduction the other. Economic systems are by their very nature cross-genealogical, while genealogical systems by definition are genealogically pure (including, as a messy extension, hybridization, chiefly confined to plants). Ecological systems attract analysis in functional terms, leading to the generation of lawlike statements of process rather than to detailed considerations of history or particularities of description of individual systems—except as exemplars (as we have noted, this is more a description of past biological practice than a proscription necessarily for the future). Genealogical systems fall largely under the purview of systematists, most of whom are occupied with tracing phylogenetic relationship through analysis of distribution of synapomorphies; the emphasis is clearly on history rather than function and causal relationships (though, of course, there are abundant exceptions—nor need this situation remain forever thus).

Continuing the comparison, we have seen that the geographic ranges of genealogical systems overlap. Geographic distributions of economic systems cannot—there can be only one system at any one place and time. It is also clear that additions and deletions to any particular system (i.e., to ecosystems on the one hand and to species on the other) are determined by wholly different sets of rules: the effects of addition or loss of component avatars to local ecosystems are determined by degree of biotic accommodation and considerations of competition. Some species can freely enter or be lost from a local ecosystem with little or no effect on the system as a whole. Other species cannot enter and become established; and some, if lost from the system, will cause the entire system great harm. Representatives of new species enter by migration; they leave by migration, or simply through death of all members present.

In contrast, the only way to add members to a species is by birth; death deletes members. But because species are organized as semi-independent "colonies" (Wright's original term), and because such colonies represent quasi-redundant packages of genetic information, there is no analogue of ecological accommodation within species. Loss of a deme in one area has no direct effect on the immediate existence of the species elsewhere—though it may represent a step towards extinction should the deme not be replaced; loss of genetic diversity as well as simple numbers of component organisms increases the probability of extinction—but only in the long run.

Such are the major differences between large-scale genealogical and ecological biotic systems. The distinction between the economic and genealogical hierarchies has a number of direct implications for evolutionary and ecological theory (for a summary of these implications from the standpoint of evolutionary theory, see Eldredge 1989, especially chapter 8). What the two systems have in common, however, is worth stressing as we turn our attention to social systems: each entity within the two hierarchies is a legitimate entity, or system. Ecosystems and species, demes and avatars—all are spatiotemporally bounded, real entities. That, at least, is our contention. It follows that colonies and social systems are likewise real. In developing the theme that social systems represent a functional reintegration of the reproductive and economic sides of organismic adaptations, it is necessary to remember that we are dealing, as an overwhelming generalization, with the local level of avatars and demes. In many organisms (including some exhibiting social behavior, such as African impalas), demes and avatars are fairly distinct. In others they are not. In other words, in seeing social systems as a reintegration, a resistance to the growing split between reproductive and economic worlds, we are not faced with reintegrating (conceptually) species and ecosystems—but the more local, restricted levels of avatars and demes. Individual social systems, like organisms, are themselves parts of ecosystems and of species.

CHAPTER SIX

The Biology
of Sociality

Organisms lead double lives. In particular, multicellular eukaryotes, and especially *animals*, lead economic lives of their own—differentiating, growing, and maintaining the soma from conception to death. During their lives, most also contribute to the economic structure of the next generation by taking part in the (generally) sexual process of reproduction. An organism's reproductive life is not strictly for its own sake, as is its economic existence: two conspecific organisms of opposite sex are usually required for reproduction; and, unlike the processes of matter-energy transfer involved in obtaining energy and building and maintaining the soma, reproduction is not necessary for the continued existence of an organism. It is a signal element of ultra-Darwinian thought that reproduction is seen as critical to the existence of individual organisms—as when competition for reproductive success is seen as a race to leave a disproportionately high percentage of one's own genetic information to the next generation. It is difficult to maintain the proposition that reproduction benefits individual reproducing organisms directly; perhaps that is why Dawkins (for example; cf. 1976, 1982) developed the argument that actually the genes themselves are competing to leave more copies of themselves to succeeding generations.

Yet as we have seen, there is no totally neat dichotomy between economic and reproductive anatomies, physiologies, and even behaviors within individual organisms of any known species—though organisms of one sex or another (most commonly male) are on occasion found to live lives almost wholly given over to reproductive functions. It is as though economics is the more general set of organismic activity. Not that reproduction is just another of a host of physiological functions comparable, for example, with respiration and excretion; but, as we have already seen, reproduction requires a healthy soma with excess energy (i.e., beyond that needed for somatic maintenance). Reproduction also, of course, relies on somatic structures dedicated either wholly or in part to the purpose. Some organisms of every species each generation do not reproduce, for a wide variety of reasons. Yet no organism is without both an economic and a reproductive life—even organisms in colonies or social systems in which labor is conspicuously divided.

As we have argued extensively in the last two chapters, it is in those systems that form as a natural consequence of the exercise of either economic or reproductive function that the dichotomy between economic and genealogical systems in biotic nature emerges. Economic systems—above the local population (avatar) level—are inherently cross-genealogical. In contrast, demes aggregate into multidemed species, themselves parts of monophyletic higher taxa, strictly as a consequence of sexual reproductive behavior. Yet there is an intermediate situation in which a less sharp distinction (at least on a case-by-case basis) is to be drawn between a local avatar and a local deme. Here, at the local-population level, the distinctions derived as outcomes from economic versus reproductive life may still be somewhat blurred. Yet the aura is one of divergence: the distinction between economic and reproductive life is almost always clearer at the local-population (deme/avatar) level than it is at the organismic level. Avatars and demes, so it seems, are best seen as intermediate (in terms of the distinctness of the processes underlying their formation) between individual organisms and local ecosystems on the one hand, and species on the other hand.

Social systems patently fit in at the local level, roughly at the level of demes and avatars. Social systems, moreover, appear to represent a distinct reversal of the ever-widening gulf between economic and reproductive activity as one goes from organism → deme/avatar → species/ecosystem. Social systems, rather, appear to represent a reintegration, a complex commingling, of economic and reproductive organismic activities. *Social systems are spatiotemporally localized entities whose existence and stability arise directly from complex interplay between the economic and reproductive concerns within and between local, conspecific organisms.*

All sexually reproducing organisms need some form of cooperative interaction for reproduction to occur (however tangential that interaction may be, as in wind-blown pollination in plants, or fertilization through simple shedding of gametes in sea water in many invertebrates). Social systems prevail within the Metazoa, and are absent within the Metaphyta,[1] because *economic* cooperation is at least a possibility in heterotrophs—and far more likely, by its very nature, than in autotrophs. Yet cooperative (as opposed to agonistic) interaction in both the economic and the reproductive spheres—without leading to the complex commingling of economic and reproductive activities that provide the cohesion of truly social systems—is imaginable within a local population of conspecifics.

Why effect a fusion between economic and reproductive organismic activities at all? Although we seem to be asking a blatantly teleological question (What is the purpose of integrating these two spheres of organismic activity? Who or what benefits?), we must agree that social systems have had an evolutionary history. And, given the vector that connects economic with reproductive activity—natural selection *is* the effect of differential economic success on (consequentially differential) reproductive success—it is only natural to seek the adaptive significance of the formation of social systems. (We travel that far, at least, with sociobiology!)

As we shall quickly see in our brief survey of different types of colonial and social systems, there is a vast difference in mode of reintegration of economic and reproductive organismic adaptations depending (to a very great degree, albeit not exclusively) on the phylogenetic relationships of the various social organisms—and on the very nature of their particular kinds of economic and reproductive adaptations. Thus, insect and homeothermic vertebrate societies (the two basic instances of socialization), while internally variable, represent two fundamentally different kinds of economic-reproductive integration—a fact only underscored by the insectlike social organization of naked mole rats (Sherman, Jarvis, and Alexander 1991), which presents a glaring exception to many of the norms of mammalian social organization (see note 6 to this chapter). And, given the very different natures of the economic-reproductive interplay in these two basic categories of social systems (i.e., insect and vertebrate), we might expect the answer to the adaptive questions—the evolutionary hows and whys underlying the appearance of social systems in the first place—to be somewhat different for each of the two. Social systems have appeared in a number of different groups; but confining our attention primarily to insects and mammals suffices to show not only that social systems have evolved more than once but also that "social systems" is a category of stable localized entity made coherent through integration of economic and reproductive organismic activities: an integration that need not occur in any one, set way.

In our view, it is the differences in the details of organization between insect and mammalian societies, rather than their superficial similarities, that provide the greatest insight into the biological nature of social systems generally. The analysis of the economic-reproductive structure of different social systems should help us to consider in greater detail why, and how, social systems of various sorts arise in the course of evolutionary history.

Colonies: Individuality and the Division of Economic and Reproductive Labor

In the context of biological systems, *division of labor* is an expression not without its ironies. There is indeed a spectrum of living systems (no doubt even richer in diversity in the past) ranging from simple single-celled prokaryotes (with no separate, distinct nucleus or intracellular organelles) through complex organisms—the metaphytes and metazoans, with billions of cells and up to two hundred different cell types arrayed in a variety of tissues, themselves organized into organs and organ systems. Midway along this spectrum are the eukaryotic cells themselves, functioning as single organisms, each with its discrete nucleus and array of cytoplasmic organelles given over to various particular economic functions.

As already seen in chapter 3, this trend to complexification is, on the surface, a trend toward subdivision of tasks—the economic from the reproductive, and the further subdivision of various aspects of the economic tasks faced by every organism. But, throughout the phylogenetic history of this increasing division of labor, a perhaps less obvious theme of amalgamation has echoed. One glaring example lies in Margulis's now widely accepted theory of symbiotic origin of the eukaryotic cell. To reiterate briefly, Margulis (e.g., 1974; cf. Margulis and Sagan 1986) has argued persuasively that organelles represent functional vestiges of phylogenetically separate organisms—organisms that had once lived free but that had taken up residence in the cytoplasm of other organisms, forever forgoing their separate existence in favor of a permanent symbiosis. The best evidence comes from mitochondria (and chloroplasts in plants), the site of cellular respiration in eukaryotic cells; these organelles have their own, separate (single-stranded) DNA that is inherited "asexually" through the maternal cytoplasm of the egg. Existence of DNA in these organelles is the strongest evidence that they were derived phylogenetically from free-living prokaryotic organisms. Other organelles in the eukaryotic cytoplasmic battery, lacking traces of independent genetic systems, are consequently less well established as separate events of phylogenetic symbiosis. There is no doubt, however, that intracellular symbiosis can occur: various species of diatoms (photosynthetic eukaryotes) shed their siliceous tests and live symbiotically in the tissues of large foraminifera (shelled ameboid single-celled eukaryotes)—their photosynthetic activities presumably acting as direct aid to the metabolic needs of the host forams (Lee and Hallock 1987).

Thus, to a significant degree the division of labor represented by the increase in cytoplasmic complexity in eukaryotes over the simpler prokaryotes represents an amalgamation of systems rather than the simple differentiation of parts. Labor is indeed divided—tasks are parceled out to discrete and semiautonomous anatomical systems—but (and herein lies the irony) the systems themselves are, in some cases at least, derived from formerly wholly separate entities rather than from differentiated subparts of an originally more homogeneous whole.

Colonial organisms enrich the perspective on the relation between economic and reproductive functions in biotic systems. Colonies such as those of the photosynthetic flagellate *Volvox*, formed by the adhesion of virtually identical cells derived asexually (shades of mitosis!) from a parental cell, which itself may be the product of either sexual or asexual ancestry, in effect represent an early stage of true multicellularity.[2] There is a polarity to the colony and evidence of a coordination in swimming activities (flagellae beat in synchronous rhythym; see Barnes 1963); and, despite the apparent similarity of the colonial cells, *Volvox* and a few of its closest relatives even evidence "a degree of cellular specialization. . . . Some of the cells are strictly somatic, while others are reproductive. The reproductive cells are larger and located away from the anterior pole" (Barnes 1963:10). Here cells are amalgamated, but there is the beginning of true differentiation from a single cell type into a series of cooperating cells that perform specialized functions—the essence of true multicellularity.

Sponges represent a further level along this spectrum of differentiation. Sponges have several different cell types, often congregated in distinct areas: some cells secrete skeletal tissue, others (collar cells) perform feeding functions, while others are for reproduction. Sponges lack true tissues, however; tissues are amalgamations of cells of a particular type, all of which contribute to the work of the tissue. Coelenterates ("cnidarians"—various jellyfish, anemones, and corals) exemplify the "tissue" grade of somatic organization; they possess true sheets of cell (tissues) devoted to various functions (feeding and respiration; reproduction). In more complex metazoans and metaphytes, of course, tissues are amalgamated into organs that perform one or more functions.

Volvox represents the kind of colony seen in its fullest extent in a single metazoan body: a collection of cells derived from a single, ancestral cell, exhibiting varying degrees of cellular (functional) differentiation But in other sorts of colonies the interplay between economic and reproductive concerns of organisms becomes more complex—and revealing. As soon as the simplest forms of true, differentiated multicellularity appeared, coloniality also appeared. If we think of *Volvox* (and, by extension if not by convention, any metazoan or metaphyte organism) as a Type I colony we immediately see that sponge and coelenterate colonies (of which there are many), not to mention colonies of true metazoans, represent another form of coloniality, distinct from the *Volvox*-metazoan spectrum of differentiated colonial multicellularity. In a very real sense, a colonial coral represents the doubling of multicelluarity—in this instance, multicellular organisms are themselves integrated into a (variably differentiated) functional whole: Type II, or "true," coloniality.

Small wonder that colonial metazoans (and metaphytes, though seemingly somewhat less so) have engendered such extensive discussion (with no hard and fast resolution) on the issue of individuality. Take, for a moment, another favorite from zoology survey courses, the hydroid coelenterate *Hydra*. *Hydra* polyp individuals live rooted to the bottoms of lakes. They feed and respire in the manner typical of all coelenterate polyps (i.e., filtering nutrients and small organic particles from the water and absorbing oxygen directly into their simple tissues by diffusion). They reproduce both sexually and asexually: sexually they shed gametes that fuse to produce a larval stage that eventually settles on the bottom and grows into a new polyp; asexually, they simply bud—a new polyp differentiates from the side of the parental polyp, eventually detaches and adopts a rooted existence nearby as a separate polyp, a separate organism capable of performing all functions, economic and reproductive, of the parental polyp. In short, asexual budding in *Hydra* produces descendant individuals as surely as does sexual reproduction.

The ontological difficulty enters when polyps fail to dissociate from the parental polyp. *Fail*, though, is a loaded word here, for all coelenterate colonies are formed through nondissociated asexual budding from a parental polyp. The result is a functioning whole, some with more obvious differentiation than others. Corals have been forming colonies for over five hundred million years (and true reefs for nearly that long as well—but reefs are cross-genealogical, ecological associations), so it is reasonable to infer that "failure to dissociate" actually means there is a functional reason underlying true (Type II) coloniality.

We need not belabor the superorganism metaphor often applied to colonies (though abjured by many zoologists in recent years, seemingly in fear of oversimplification of colonial biology). But what, indeed, does constitute the individual in a colony of meta-zoans formed by asexual budding? Each coral polyp is a distinct organism (as are separate *Hydra* polyps), yet the colony is also a distinct entity unto itself. The colony, for one thing, is a clone—all its component polyps share exactly the same genome, save accumu-lated mutations. In an ecological context (such as the reefs many coral colonies find themselves in), it is the colony itself that does the interactive work: providing shelter in its interstitial niches; having a mass, blanket effect on the dissolved particulate and chemical com-ponents of the surrounding sea water; and providing sustenance for predators who graze on coelenterate polyps en masse. The colony itself is indeed an entity, an individual. There are simply two levels of individuality, polyps and entire colony, in such situations.[3]

It is when individual polyps (or, more generally, organismic com-ponents) of colonies begin to lose their identity as separate organ-

isms (i.e., capable of performing all vital economic, as well as reproductive, functions) that colonies themselves emerge as *the* functional, holistic entity. Division of labor is at the cost of independent
capabilities of the component organisms. Gone, in many metazoan
colonies, is the redundancy of the simplest of colonies, in which
each organism (polyp or even cell, in the case of Type I colonies) is
genetically as well as functionally the same as all the rest.

Siphonophorans are an arresting case in point. Hydrozoan relatives of the aforementioned *Hydra,* siphonophorans form large and
complexly differentiated floating jellyfish colonies: *Physalia,* the Por-
tuguese man-of-war, and *Velella,* the purple sail, are two of the
better-known genera. Each colony is an amalgam of medusae (floating individuals) and polypoid individuals. In some siphonophorans,
specialized medusoid individuals form pulsating bells, providing
thrust for locomotion of the entire colony. Polypoids are differentiated typically into separate feeding, defensive, and reproductive
entities. No swimmer encountering such a colony, however, has any
difficulty perceiving it as a single individual; indeed, it is difficult, in
contemplating the division of labor of such a complex coelenterate
colony, to see it as different in any way from a typical metazoan
individual organism—except, of course, that each component part
is itself a highly derived version of what was once (phylogenetically)
a separate, independent, economic and reproductive organism. But
just as the individual cells of a metazoan surrender their autonomy
in becoming specialized within a larger soma, so do differentiated
metazoan organisms: a *Physalia* is a repeat performance of a theme
already established to produce a (cellularly) differentiated hydrozoan polyp individual in the first place.

Bryozoan colonies illustrate this theme of double differentiation
even more graphically than any hydroid coelenterate ever could.
Bryozoans (ectoprocts, or true bryozoans) are lophophorates, usually considered to sit astride, or nearly so, the great branch point of
metazoan phylogeny that saw protostomes (annelid worms, mollusks, arthropods, and others) and deuterostomes (echinoderms,
chordates, and others) go their separate evolutionary ways. Lophophorates are coelomate metazoans in the fullest sense: those that
live as single individuals (such as phoronid "worms") have fully
differentiated feeding and digestive systems, excretory organs, mus-
cle and circulatory systems, and reproductive organs. Bryozoans, in
contrast, live almost exclusively in colonies. Each individual (zooid)
within the colony is tiny and is able to get by without discrete

138 THE BIOLOGY OF SOCIALITY

respiratory, excretory, and circulatory systems (Barnes 1963)—a reflection, it is presumed, of their microscopic size. The rest of the complexly differentiated, multicellular lophophorate anatomy is very much present, however, in each bryozoan zooid.

Bryozoan colonies typically display an astonishing degree of differentiation of the zooids. Typically, most zooids (autozooids) are feeding individuals, equipped with a fully functional lophophore (tentaclelike system). Other zooids (collectively, heterozooids) are modified, often radically so, for defense (avicularia; cf. Winston 1986), cleaning (vibracula), and reproduction (some zooids become modified for egg production and brooding). There is clearly a high degree of neurological and behavioral integration among the zooids of a bryozoan colony.

Just as in the siphonophoran case—and all other colonial examples yet known, for that matter—division of labor (both between economic and reproductive functions and within economic functions per se) is a matter of differentiation of colonial components to perform particular tasks. There are two important aspects to this general characterization of mode of specialization of function within all true colonies. First of all, such cooperative division of labor inevitably results, as a side effect, in the formation of a larger-scale (i.e., than the component separate organisms) functional entity that itself has a beginning, a history, and an end—an entity, moreover, that itself engages in reproductive activities. (Colonies make more colonies.) It is entirely appropriate, in that case, to examine the nature of such entities, and to explore the nature of their interactions with other abiotic and biotic entities.

Specializing in one function, relaxing the need to perform many, is presumably a benefit to the individual as well as to the group as a whole. But the second general aspect of coloniality deeply colors any interpretation of the significance of division of labor within colonies: the fact that all colonies (as all metazoan bodies) are derived from asexual reproduction of component parts, starting from an ancestral entity. Genetic (near-) identity among thousands, or even millions, of integrated colonial components both fosters the impression that a colony is an individual in some truly meaningful biological sense and allows us to understand how such cooperation could become established and further refined in the first place. The subdivision of labor is, genetically speaking, for the enhancement of the economic and reproductive activities of one entity—the colonial phenome/genome. In other words, division of labor in a siphonophoran colony than no more remarkable (indeed, in genetic terms, it is exactly the same thing) than is the differentiation of a human body into various "vital" and reproductive systems. All systems work in concert to keep the one single entity (the individual, with its own genome) alive and reproducing.

Sociobiology, it seems fair to say, had its inception (at least empirically) in the study of insect societies. It is now perhaps obvious that insect social systems are very like colonial systems—even to the point that every component organismic part of an insect society is related to every other, albeit not in quite as straightforward a manner as in an ectoproct colony. Patterns of division of labor in insects are very much in the same mode as in typical colony systems; and much the same arguments that have been made concerning the significance of the genetic (near-) identity shared by all members of an insect society are evident with all marine invertebrate colonial systems.

Insect social systems, as we shall soon see, represent a banding together of separate insects to work together; but it is more appropriate to see insect societies as a differentiation of individual types within a descent group—just as we have seen is the case in metazoan colonies. Mammalian and avian social systems (with but one exception known to us) are something entirely different: almost invariably cross-genealogical (i.e., on a within-species microscale), vertebrate social systems seem to represent a true fusion and integration of economic and reproductive adaptations that deviate significantly from the colonial mode of integration and division of labor between economic and reproductive activities and among economic functions.

Reproductive Structure and Social Systems

In the landmark work that gave sociobiology its very identity, Wilson (1975) wrote of the "four pinnacles of social evolution." To Wilson, colonial invertebrates form societies every bit in the same sense as do insects and human and nonhuman vertebrates (the remaining three of his four "pinnacles").

Although there is little point quibbling over the status of invertebrate colonies (of either type discussed above) as societies, it is curious that more has not been made in sociobiological literature of the obvious fact that insect societies are organized far more like (metazoan) invertebrate colonies than like vertebrate social systems. Two points seem especially critical: (1) the manner in which labor (both economic and reproductive) is divided within colonial and insect social systems differs drastically from the usual situation among vertebrates (a point considered in detail in the following section); and (2) insect societies display patterns of enhanced relatedness among components, which, if not at the level of clonal identity displayed in marine invertebrate colonial systems, has nonetheless a great deal to do with their organization. Indeed, considerations of haplodiploidy in hymenopterans (with consequent difference in degrees of relatedness among and between brothers and sisters, and parents and offspring) led directly to Hamilton's (1964a, 1964b) predictions of degrees of selfishness displayed by members of each sex within a hive—and to his theory of inclusive fitness. It is no less obvious that degree and nature of kinship varies among vertebrate societies, but especially between vertebrate societies generally and insect social systems. At the very least, hymenopteran haplodiploidy, especially where a single queen is laying the eggs for a generation, bespeaks a kinship pattern more reminiscent of clonal colonies than of vertebrate social systems, which, beyond small family units, are almost invariably cross-genealogical in composition.

The literature on animal sociobiology, including "socioecology," is heavily (though certainly not exclusively) devoted to considerations of reproductive biology. Eusociality in insects, for example, is defined as "(1) cooperation in caring for the young; (2) overlap of at least two generations capable of contributing to colony labor; and (3) reproductive division of labor, with a mostly or wholly sterile worker caste working on behalf of individuals engaged in reproduction" (Oster and Wilson 1978:3). Sociality, in other words, is all about reproduction.[4] Nor is the emphasis on reproductive biology restricted by any means to considerations of the social insects—most of the essays in the collection edited by Rubenstein and Wrangham (1986) focus on reproductive strategies within various avian and mammalian social systems. Hamilton's notion of *inclusive fitness*, coupled with the Maynard Smith notion of *evolutionary stable strategies* (whereby stability in social systems is achieved through intersection of separate male and female reproductive interests), has long since been extended to vertebrates—despite the obvious differences between (social, especially hymenopteran) insect and vertebrate modes of reproduction and inheritance.

Sociobiologists, of course, are fully cognizant of the fact that insect and vertebrate reproductive systems differ markedly, even drastically. The extension of kin selection theory to vertebrates is not at all a straightforward matter. In the case of clonal colonial organisms, division of economic (and reproductive) labor among clonal elements can easily be read, in genetic terms, as "all for one, one for all." Benefits accrue, too, to the individual specialized polyp or zooid—as long as someone else is taking care of the other tasks that normally a single, independent organism would be compelled to carry out.[5]

Social insects, of course, are not quite like marine colonial invertebrates. Genetic identity does not obtain among the elements; rather, all hymenopterans within a colony are related, but to varying degrees. It makes sense, then, to see the activities of various castes (and the fact that individuals will change roles as they progress through their lives; Oster and Wilson 1978) as devoted to the support of reproductive activities in proportion to the relatedness of the individuals. It is not quite, in reproductive terms, "all for one, one for all" with the eusocial insects. Yet all members of a hive are related, and the argument that social behavior hinges to a great degree on fitness is, if not as straightforward as it is with clonal marine colonial invertebrates, logical. Cooperation—subdivision of economic labor, its separation to a large degree from reproductive labor, and the application of economic labor directly to reproductive activities (as in the feeding of larvae)—has, at least, the effect of propagation of the underlying shared genes (and apparently largely in proportion to degree of genetic relatedness, as Hamilton's theory predicts).

Be that as it may, to whatever extent genetic relatedness differs between clonal invertebrates and colonial insects, the two are genealogically far more coherently pure than virtually any vertebrate social system. Vertebrate social systems—beyond stable pair-bonds and some instances of cooperative breeding found in some species—are invariably cross-genealogical.[6] Application of kin selection theory to vertebrates is accordingly restricted to situations where older sibs or extended kin (with appropriately calculated lesser proportions of alleles in common) share in aspects of the reproductive process. Some biologists (e.g., Zahavi 1974) have openly doubted whether kin selection is ever appropriately ascribed to vertebrates. And one of the most prominently cited cases (cooperative breeding in Florida scrub jays) is, according to those who performed the study, best understood as a strategy to maximize individual reproductive success rather than as a straightforward case of kin selection (Woolfenden and Fitzpatrick 1984). We discuss the Florida scrub jay social system in greater detail later in this chapter.

Yet vertebrate social systems commonly are complexly hierarchical in organization. Localized kin systems, no matter what their size, are usually parts of larger systems; Moss (1988) has described the hierarchical arrangement of local herd systems of African elephants, in which several mother elephants and their children of various ages constitute a single social system.[7] The larger systems into which pair-bonds or larger kin groups are integrated are cross-genealogical—there is no systematic pattern of relatedness that can be used to predict patterns of cooperative or competitive interactive behavior. To be sure, a greater proportion of alleles is likely to be shared within local breeding populations than between such demes in a species, but the lack of a coherent pattern of relatedness (i.e., a means of telling which organism shares more alleles with which) limits the applicability of kin selection theory to vertebrates.

Yet the theme that social organization in vertebrates nonetheless hinges on considerations of relative reproductive success is symbolized by the impact that Maynard Smith's notion of evolutionary stable strategies has had on vertebrate sociobiology. Vertebrate social systems are often described as though they are nothing more (nor less) than complex mating and reproductive systems. All vertebrate social systems can be described in terms of three possibilities (cf. Maynard Smith 1977): care of young (1) by both male and female, (2) by one sex and not the other, or (3) by neither. As Rubenstein and Wrangham (1986:5) have put it: "Whether there is to be parental care and, if so, which parent is to provide it depends on the relative effectiveness of uni- or biparental care, the likelihood of a deserting partner finding an additional mate, and the extent to which a female's ability to provide care reduces her ability to produce future offspring."

It is to be stressed that this analytic system is addressed specifically to predicting (or simply understanding) what mating system is to be expected, given certain aspects of the biology of the organisms in question. Yet the concept of the evolutionary stable strategy is widely cited as one of the three basic linchpins of sociobiology. To a marked degree, sociobiology sees social structure as synonomyous with reproductive biology—and social behavior as both a consequence of underlying genetic structure and a determinant (via selection) of future allelic frequencies (which is to say, evolution). Social

structure is all about reproduction: reproductive strategies determine social structure. The very fact that Maynard Smith dubbed his
reproductive strategy/analytic rubric *"evolutionary* stable strategy"
demonstrates the close conceptual connection between these various
elements. The syllogism goes something like this: evolution *is* natural selection *is* differential reproductive success; social systems have
evolutionary histories; ergo the reproductive systems, found to differ systematically among (vertebrate) social systems, are fundamental to social systems.

The connection, in other words, between social organization and
reproductive biology is dictated by assumptions held a priori about
the nature of the evolutionary process. There is little doubt that
there are tight connections between economic success and reproductive success and that differential reproductive success biases genetic representation in the succeeding generation—yielding evolution through natural selection. But there is a world of difference
between a statistical *effect* of relative economic success on reproductive success, and the view that the very fabric of life is organized
around a process that determines what alleles will be represented in
the following generation.

Our view is quite different: we, along with some biologists, see
both economic and reproductive advantages accruing to social organisms. As in the biotic world generally, relative economic success
may be reflected in relative reproductive success (natural selection)—and the alleles underlying the favorable variations may show
up in the evolutionary ledger as a side effect of the variation in
economic (and reproductive) behavior. *Allelic representation in suc-*
ceeding generations is not the cause, but the effect, of such differential
economic, and hence reproductive, success. At least in vertebrates, the
actual organization of biotic systems—social systems in particular—
should be read first and foremost as a simple, moment-by-moment
maximization primarily of the economic lives of component organisms. Economic as well as reproductive adaptations are developed
so as to be utilized in conjunction (at least in large measure) with the
activities of other local conspecific organisms. Social fabric arises
from the moment-by-moment interactions of local conspecifics in
pursuit of the economic and reproductive aspects of their lives.

In most sociobiological analyses, true to the ultimate theoretical
base of sociobiology, economic behavior is subordinate to reproductive: economic behavior is, ultimately, *for* reproduction (precisely
the point, for single organisms within demes generally, of Dawkins's

"selfish gene"). This point of view is most persuasive in the truly clonal colonial situation—which, we have argued, is no different in principle from that of a single multicellular organism: general economic success is necessary for reproductive success, and all (genetically identical) parts, no matter what their particular economic or explicitly reproductive tasks, can at least be imagined as sharing equally in "concern" for the fate of their mutually shared genetic information. The situation is a bit more exiguous in insect societies, which, while not clones, at least exhibit a spectrum of degrees of genetic relatedness within "monophyletic" genealogical lineages. Vertebrates, however, as the overwhelming rule, quickly abandon genealogical homogeneity immediately above the restricted local kin group—itself generally a part of a larger, inherently cross-genealogical system. Better, especially with vertebrates, to fall back on the more open, non- ultra-Darwinian view of organismic organization and function, taking neither economics nor reproduction as inherently the more general or fundamental aspect of an organism's life, and see what implications the view has for understanding the biology—the structure, ongoing function, *and* history—of social systems.

Economics, Division of Labor, and Insect Social Systems

Eusocial insects and marine invertebrate colonies share a second fundamental similarity (i.e., beyond genealogical integrity or near-integrity) that sets both off from typical vertebrate social systems. That similarity resides in their mode of division of both economic and reproductive labor.

The colonial/eusocial insect mode of labor division is basically a simple parceling of tasks—a system of task assignment, moreover, that counts economic and reproductive tasks together. Some polyps, zooids, or individual bees are assigned either purely reproductive or purely economic tasks. In the case of eusocial insects, the picture is complicated by the presence of helper castes that supply economic needs to larvae—the application of economic work not to the economic needs of the organism itself but to offspring. Such labor is, of course, correctly perceived as reproductive in nature—and the fact that helper castes are sterile, that is, contributing to the genetic fitness of other members of the social system (an apparently altruistic act), is what prompted Hamilton to propose his concept of kin selection (based on the presence of the haplodiploid system in eusocial insects).

Wilson (1985) has recently reviewed the "ontogeny" of tasks commonly encountered, especially in more advanced, complex, eusocial insect societies. Not only are there separate physical castes, but an individual worker may pass through "a sequence of temporal castes as it passes through its life-span" (Wilson 1985:1490). Wilson cites, as an example, the sequence of a worker tending to the queen or larvae, then participating in nest building, and finally foraging for food outside the nest. Tasks are parceled out to different physical castes, but even within physical castes, task assignment can vary significantly over the lifetime of a single organism. Nor is reproduction exempt from flexibility in assignment: even where there is but a single queen, her loss to the colony immediately prompts another female, by definition from another caste, to attain reproductive status—as occurs not only in eusocial insects, but also in naked mole rats.

Thus the caste system is not written in stone. Yet the conditions that cause genetically similar organisms to develop physically as a member of one or another caste are fairly well known; basically, as a eusocial insect develops, its options decrease (much as cell potentiality decreases in ontogeny—as Wilson [1985] has pointed out). The division of labor within eusocial insects, in fact, bears a striking resemblance to the differentiation of genetically identical cells within a metazoan body. Much as separate economic functions are handled by various organelles within the cytoplasm of eukaryotic cells, the cells of a complex (e.g., coelomate) metazoan are differentiated into as many as two hundred different types. An organ system, specialized for a single (albeit complex) task, is typically composed of a number of different tissues, each with cells specialized to one (or, at least, a restricted subset) of the subtasks of the organ system's generalized task. And, even though cell types may change their roles somewhat during their "lifetimes," and even though, indeed, an entire metazoan soma develops from a single omnipotential cell, the basic theme in the metazoan soma is a rather strict and formal cell/tissue/organ compartmentalization of tasks: a compartmentalization between economic and reproductive tasks, and, within each major category, specific subtasks.

So, too, with colonies and eusocial insects. Tasks that are fundamentally economic in nature are subdivided among physical castes.

And despite the temporal division of labor (i.e., within the lifetimes of single workers) that Wilson (1985) has discussed, the point is that there is (1) an ontogenetic trajectory of tasks and (2) a fixed subset of tasks a worker might encounter through its lifetime—no single worker is likely, or able, to assume any task at any given time, let alone engage in the entire spectrum of economic and reproductive tasks that exist in the colony. Rather, everyone has his or her job, and despite some temporal changes in job assignment, the overall work of the eusocial insect colony (as in marine invertebrate clonal colonies) is done by several or many different kinds of specialists all carrying out a subset of the work—the economic and reproductive work.

We have argued that first a fundamental dichotomy exists between economic and reproductive labor, followed by subdivisions of tasks within both these general categories. This holds for the eukaryotic cell, metazoan bodies, marine clonal colonies, and insect eusocial systems. Sociobiologists, as we have already acknowledged, tend to view much of the economic activity of workers as fundamentally reproductive; when a worker feeds or otherwise cares for a pupa or larva, that behavior is, in a very real sense, reproductive in nature (regardless of the genetic relatedness of adult to immature individuals). Switching to vertebrates for a moment, a female fox standing guard, ready to attack to defend her kits, is risking her own life for the sake of the kits' safety. Her behavior is fraught with both economic and reproductive significance.[8] Thus, we must ask, is there really a fundamental distinction to be drawn between reproductive and economic activities within the eusocial insects?

We reiterate our earlier point that the distinction between economic and reproductive activities is blurred by the realization that there is an economic cost to reproduction: all physiological activity requires energy, to be supplied in all cases through the economic activities of the organism. Energy (as well as physical cells, tissues, and organ systems) must be allocated to the reproductive process. Lack of sufficient energy commonly (at least among animals—some plants seem to constitute an exception) results in cessation of reproductive activities first, reproduction being the one physiological activity unnecessary to the actual physical existence of an organism. There are, it seems, important distinctions to be made between reproductive and all other activities engaged in by organisms: one (reproduction) requires the other, but the converse is not true; the two sets of needs (economic and reproductive) of organisms are often in conflict (see note 8); and the consequences of there being two sets of activity only become starkly clear when we examine groups that are formed as a result of these two classes of activity (genealogical entities such as demes, species, etc., and economic groups such as avatars and local ecosystems).

As for organisms, so for clonal marine and eusocial insect colonies. Reproductive polyps in an ectoproct colony require nourishment to perform their functions—every bit as much as do vibracula and avicularia, which are providing defensive and cleansing (but not feeding) economic functions. Worker ants feeding larvae supply the connection between the economic and reproductive functions already familiar in the workings of any organism. The distinction between economic and reproductive functions is no more blurred, as a consequence, within these complex systems than it is within single organisms when energy (and soma) are utilized for reproductive purposes. Indeed, the transfer of energy from economic workers to reproductive caste members simply strengthens the analogy between single, multicellular organisms (even single-celled eukaryotes) and colonial and social systems.

Colonies and eusocial insect systems appear at first glance to represent a fusion, an alliance, between specialist organisms, reintegrating separate reproductive and eusocial functions that otherwise tend to form distinctly economic and reproductive entities. But such is emphatically not the case: phylogenetically, each of the eleven or so (Wilson 1985) known lineages of extant hymenopterans that developed sociality sprang from nonsocial taxa. Phylogenetic development of sociality appears to have been a process very much like the evolutionary development of clonal colonies and the derivation of multicellularity from single-celled organisms. This process begins with association of "stand-alone" organisms—that is, organisms capable of performing the full gamut of economic and reproductive functions. There follows a phylogenetic accumulation of division of labor along the lines sketched very generally above: reproductive versus economic, and many subsets, especially of the economic—plus the category "economic activity with reproductive purpose." It is now notorious that there exists, within several well-studied lineages, a gradient between very primitive social systems and extremely complex ones.

Thus, in a sense, insect social systems represent not so much the fusion of many organisms into a single system as the differentiation of kinds of organisms (within the same species) that perform certain narrowly specified tasks (which may allow many organisms—millions, in fact—to live cooperatively alongside one another within huge social systems). But it is also possible, and perhaps more accurate, to view these systems as a true fusion of the economic and reproductive activities of separate organisms into a joint economic/reproductive community that does blur a distinction: the distinction between local economic and reproductive communities—avatars and demes.

Eusocial insects, then, share with the differentiated cells of the metazoan body, and with the differentiation of individuals within marine clonal colonial systems, two fundamental characteristics: (1) "monophyly" or nearly so of descent of components—that is, genetic identity (through mitotic cell division or asexual budding) or near-identity (through insect haplodiploidy and descent from single queen); and (2) division between economic and reproductive labor with further subdivision, especially in economics, as well as activities that connect the economic with the reproductive (reproduction requiring energy and a physical contact without which it simply cannot take place). These similarities transcend the status of mere analogy: they indicate the fundamental likeness in basic nature between all these categories of systems, and they bear directly on the issue of the fundamental nature, hence mode of origin, of (insect eusocial) social systems.

As we have seen, sociobiologists routinely *define* social systems as organizations characterized in terms of certain reproductive properties of component organisms: overlapping generations, parental care, and the existence of "more or less *non*-reproductive workers or helpers" (Andersson 1984:165). This is what social systems purportedly *are*—and are therefore, it is presumed, designed to be. In this paradigm, social behavior is about reproduction; the ultra-Darwinian input (which served as the very impetus to found sociobiology) emphasizes differential reproductive success as the cornerstone of the evolutionary process. Social systems posed a now-classic paradox: how to explain in Darwinian terms (Darwin himself was aware of the problem) the existence of sterile castes so selflessly devoted to the care and feeding of offspring not their own. It was Hamilton's contribution to explain this apparently altruistic behavior as an extended form of genetic selfishness: the degree of altruistic helping was predicted (and later confirmed) to be proportional to the degree of genetic relatedness among members of the colony. Thus the origin of the powerful concept of kin selection. It is also worth noting that the postulate that altruism is really a form of competition in disguise constitutes the key insight that allowed an ultra-Darwinian interpretation of social behavior to be formulated in the first place. Of course, viewing cooperation solely as a form of competition obscures the plausible and certainly more simple alternative postulate: cooperation in social systems may be to the immediate economic advantage of the organisms, regardless of their degree of genetic relatedness.

But there is a distinction to be drawn between the prediction and demonstration of a particular phenomenon of reproductive behavior (in this instance, of economic behavior expended on nourishing of young) and the overarching assumption that somehow colonial, social behavior is to be construed as essentially reproductive—as if, in other words, all behavior, certainly including economic, were to be understood as being carried out strictly for reproductive purposes. Yet such is the assumption underlying a characterization of social systems strictly in reproductive terms. Here is where the analogy between insect eusocial, marine clonal colonial systems, and the organization of the metazoan organism is critical. If indeed it is true, as we have endeavored to establish, that the fundamentals of organization are the same in these different kinds of systems, we must ask: are asocial metazoans, or clonal colonies, likewise to be construed as primarily reproductive entities?

Our review of reproductive and economic systems in previous chapters strongly suggests that it is impossible to maintain that an organism is more an economic than a reproductive entity, or vice versa. Although Dawkins (e.g., 1976, 1982) has come close to asserting that organisms are mere vehicles for genes and that the entire purpose of an organism's existence is to house and provide energy for the reproduction of its machinating genes, it must still be conceded that those genes are in every somatic cell as well as in the germ line. Organisms are no more "about" reproduction (the transfer of genetic information in the making of more organisms of like kind) than they are purely matter-energy transfer machines (though, if one were insisting on a single description, the notion that organisms are matter-energy transferring entities—temporarily formed "dissipative structures"—that happen also to reproduce may be more accurate than the alternative touted by ultra-Darwinians: organisms are reproductive machines that happen to require matter-energy process to carry on their activities).[9]

And thus our basic point about insect eusocial systems: even though it may seem compelling to define such systems purely in reproductive terms, we cannot—because they are fundamentally so like individual metazoan organisms and clonal marine colonies. These latter entities clearly do not exist purely for the sake of their reproductive functions, and we must also see eusocial insect societies as fully balanced economic-reproductive cooperatives and *not* merely as reproductive cooperatives with a necessary economic gloss. In some contexts, as we have seen, Wilson and other important contributors to sociobiology have indeed acknowledged direct, immediate economic benefits to social existence; yet the explanation, in causal terms, for the origin of such systems immediately reverts to a reliance on reproductive advantages—as if the only functional aspect explaining the existence of a social system were reproduction. Insect social systems are no more purely, or even primarily, about reproduction than any organism, most certainly including ourselves, is strictly about reproduction. Moreover, anyone contemplating the ecological work of, for example, an ant colony will immediately concede that as an entity, social systems are not just reproductive communities. Species, we have argued, do no economic work: they arise purely from the reproductive behaviors of organisms. Similarly, colonies and social systems cannot arise purely as outgrowths of reproductive behavior and be so clearly ecological interactors at the same time. (We pursue this topic of social systems as large-scale entities further below.)

Economics, Reproduction, and Vertebrate Social Systems

However we view insect societies—as differentiated colonial systems like clonal invertebrates (and in turn like metazoan body systems) or as integrations of the separate economic and reproductive aspects of asocial insect forerunners—we must immediately acknowledge that vertebrates, in their social systems, develop a radically different mode of subdivision of labor. Though lek behavior (in which male reproductive activity at any moment, or within any breeding season, is concentrated in a very few dominant males) has overtones of a reproductive caste system, mammals and birds more generally do not divide labor in a manner reminiscent of insects, marine invertebrate colonial systems, or, for that matter, a metazoan body. Generally speaking, physiologically reproductively active males and females engage in reproductive activities; and all organisms of both sexes engage in independent economic activity—a generalization that demands closer scrutiny, especially in the case of humans. At the very least, there is far less desertion of individual autonomy in vertebrate social systems than in eusocial insects (though there are those naked mole rats). It could be argued that vertebrate systems are merely less differentiated than insect systems and that the haplodiploidy of Hymenoptera in particular has a different set of (genetic) implications from the normal diploidy of vertebrates— circumstances that encourage, for example, the establishment of sterile castes in insects but not in vertebrates.

Yet the modes of complex interaction between economic and reproductive concerns of vertebrate organisms convey a very different aspect to vertebrate vis-à-vis insect social systems. The very fact that vertebrate social organisms remain independent organisms to a very great degree implies a rich field of possible interactions between economic and reproductive activities. Then, too, because vertebrate organisms do not differentiate into specialized economic or reproductive caste members, and because each organism is living an economic and reproductive life much as if it were a member of an asocial species, the interlocking of reproductive and economic activities in social vertebrates amounts to a fusion, on the local-population level, of the separate avatars and demes formed by asocial organisms. *Vertebrate social systems are local populations that arise from the intertwined economic and reproductive activities of their component organisms.* Above that local level, of course, the basic dichotomy remains: social systems are parts, simultaneously, of local ecosystems *and* of species. Nothing can stave off, permanently, the biotic consequences of the division between economic and reproductive activity.

Because, as a rule, each vertebrate social organism is simultaneously a fully economic and reproductive organism, the division of labor within vertebrate social systems does not conform to a primal reproductive/economic dichotomy. Indeed, the interactive "glue" that binds vertebrate social systems stems in large measure from the intricacies of interaction between economic and reproductive activities between organisms. Polyps or zooids in clonal marine colonies, not to mention the tissue-forming cells of the metazoan and metaphyte body, are physically conjoined. Interaction is mediated by direct physical connection and often further facilitated by neural, circulatory, and skeletal connections among components. Social systems, consisting of separate organisms, are coherent instead by virtue of behavioral (including sporadic physical) interactions. Specifically, there are economic consequences of reproductive behavior—and vice versa—in all social systems (including those of insects).

Biology of Sociality in Florida Scrub Jays

Florida scrub jays at first appear to contradict the general proposition that social systems represent fusions of economic and reproductive organismic activities. The reason is that they barely constitute social systems at all: according to Woolfenden and Fitzpatrick (1984; our sole source for this example), Florida scrub jay sociality is rudimentary. Sociality in these birds is said to be manifested strictly around the phenomenon of cooperative breeding. Yet the example contains a number of relevant points.

As we have already remarked, sociobiologists have often cited Florida scrub jay cooperative breeding as an example of kin selection in vertebrates—a conclusion expressly rejected by Woolfenden and Fitzpatrick (1984) as the result of their analysis of the data they collected during their decade-long study. Florida scrub jays constitute a relict aggregation, separated by some 1600 km from the main concentration of the species in western North America. Significantly, western scrub jays do not engage in cooperative breeding.

Florida scrub jays are restricted to scrub oak habitat, which has an exceedingly patchy distribution. Woolfenden and Fitzpatrick (1984) conclude that the dense populations of scrub jays wherever suitable habitat is developed reflect a level at or near the maximum carrying capacity of the environment. They note, further, that distributions are sharply limited by the unavailability of marginal habitat: blue jays quickly replace scrub jays in all but pure scrub oak habitat.

Florida scrub jays form permanently monogamous pair-bonds that usually terminate only upon the death of one of the mates. Woolfenden and Fitzpatrick document a "divorce" rate of only 5 percent. The signal element underlying scrub jay social behavior is the phenomenon of delayed breeding: rather than leaving the nest within a year's time to establish territories, pair, and mate (as, for example, occurs in western scrub jay populations), Florida scrub jay offspring typically become "helpers" through "site tenacity"—offspring tend to remain within parental territory.

According to Woolfenden and Fitzpatrick (1984), Florida scrub jay helpers are most often offspring of the mating pair whose reproductive efforts they are aiding. Over 90 percent of helpers are the offspring of at least one of the mating pair. Yet helping continues when the parents are replaced by another pair-bond: 10 percent of helpers are aiding the reproductive efforts of nonrelatives.

Helpers cooperate in all phases of reproductive effort save, of course, mating, egg laying, and (interestingly) nest construction and incubation. The helpers' activities include foraging, feeding nestlings and young fledglings, and territorial defense. According to Woolfenden and Fitzpatrick (1984:83), "all foraging, resting, roosting, territorial, and antipredator behavior (i.e., of helpers) qualitatively resembles that of the breeders. Group members spend much of the day in close proximity to one another, and helpers often participate in sentinel activity while the rest of the group forages." Only 50 percent of breeding pairs during any one breeding season have helpers.

Crucial to a consideration of the biological basis of scrub jay sociality is the fact that these birds *"always* reside in territories with well-defined boundaries defended *year round by all group members"* (Woolfenden and Fitzpatrick 1984:101; all italics ours). These territories are very stable, remaining essentially "the same piece of ground for many years." Most critical of all, "ownership of territories is passed on through sequential mate replacements *or through inheritance by helpers"* (Woolfenden and Fitzpatrick 1984:101). Woolfenden and Fitzpatrick (1984:101) conclude that *"the territory represents a land bank, the defense of which is vital to ongoing reproduction by the breeders, and to survival and later reproduction by the helpers."*

Helpers become breeders in several ways, most of which involve obtaining territory in one way or another by inheritance. Sometimes they replace mates outside their parents' territory, or (rarely) they may establish new territories between preexisting ones; but, more usually, they inherit all or a subdivided portion of the territory in which they were helpers (i.e., usually, but not invariably, their natal territory). (Most females disperse to new territories; Woolfenden and Fitzpatrick (1984) report that roughly 56 percent of breeding males inherit territory.)

Crucial to the interpretation of the purely reproductive significance of Florida scrub jay behavior is the observation that "pairs with helpers fledge 1.5 times more young than do pairs without helpers." Moreover, the data suggest that territory quality is less important than the presence of helpers in fledgling production. Helpers, moreover, tend to produce more offspring of their own (i.e., once they become breeders) than do nonhelpers—but the longer they act as helpers, the lower are their survival rates once they become breeders (for unknown reasons). Woolfenden and Fitzpatrick (1984, ch. 8) conclude that a young jay's inclusive fitness is increased more by breeding than by acting as a helper.

Thus Woolfenden and Fitzpatrick (1984) conclude that the helper behavior of Florida scrub jays represents a delaying tactic that is used until a "breeding vacancy" can be located outside the territory or created within it. Unlike the situation in western North America, where there are many habitats suitable for scrub jays, available habitat for Florida scrub jays is severely limited. Woolfenden and Fitzpatrick point out that biologists have often failed to look at total numbers of successfully produced offspring over an organism's entire reproductive lifetime—which can reflect the input of a number of different variables. They conclude (Woolfenden and Fitzpatrick 1984, ch. 10) that the cooperative-breeder behavior of Florida scrub jays reflects far more a strategy for maximizing fitness through the actual reproductive efforts of helpers-turned-breeders than it reflects the action of kin selection—in which helper jays are hypothesized to be contributing to their inclusive fitness by helping to rear (predominantly) younger sibs. It's the territory they are after, so that they can spread their own genes, not simply those they share with their sibs.

The territory does indeed seem to lie at the heart of the matter. Woolfenden and Fitzpatrick (1984) are content to refer to these stable and partly heritable (obviously *not* in a genetics sense!) territories as primarily reproductive in purpose—as in their statement, cited above, that "the territory represents a land bank, the defense of which is vital to ongoing reproduction by the breeders, and to survival and later reproduction by the breeders." It is as if, in their model, breeders remain in the territory, defending it year-round, only to insure its availability for the *next* breeding season. Yet helpers defend the territory for *survival* as well as reproduction (survival, presumably, in order to have a chance to reproduce). Even if kin selection need not be invoked to explain the pattern, the phenomenon of sociality arising from cooperative-breeding behavior is still seen as arising strictly from the reproductive imperative—the (competitive) need to maximize genetic representation in the next generation.

Yet territories are, foremost, all about food availability, protection from predation, and all the other ingredients of the economic life—of maintaining one's existence. The breeding adults are surviving as much as are their helpers as they cooperate the year around to defend the territorial boundaries. Particularly in view of the putative near-maximization of scrub jays in available Florida scrub oak habitat, the definition and defense of well-demarcated territories is as much about pure economics as it is about reproduction. We may well conclude that helpers are cooperating in territorial defense because they want the territory—but that they want that territory as much to survive as to reproduce.

Looking at Florida scrub jay behavior is very much like considering the notion of natural selection itself. As we set forth earlier (chapter 3), natural selection is the differential effect on reproductive success wrought by differential economic success. Yet modern evolutionary theory is generally content to define natural selection purely as differential reproductive success—and this important dynamic in nature devolves simply to a race between local conspecifics to leave relatively more copies of their genes to the succeeding generation. Only when we concede that there are economic as well as reproductive aspects to an organism's life can we see the interactions between them (at various different levels). In assessing Florida scrub jay sociality, asserting that the vastly longer period of the nonbreeding vis-à-vis the breeding season is irrelevant or is to be construed solely as a waiting and staging arena for the next breeding season, is to ignore perhaps the primary function of the behavior itself: the delineation and maintenance of economic resource space.

Sex and Economics in Great Ape Social Systems

The literature on primate social behavior typically places strong emphasis on economic activity. For example, among various lemur species, mating is rarely observed (as is the case for most organisms). Most behavior accessible to systematic observation is economic— especially feeding (see, e.g., Richard 1985; Tattersall 1982, ch. 7).

The four great apes differ rather markedly in the details of their social organization. Orangs are essentially solitary, while gorillas form stable groups consisting of one or two adult males and generally no more than six females. Chimpanzees and bonobos (pygmy chimpanzees) form much larger communities, yet the details of "mating systems and association patterns" differ substantially between these two closely related species (Wrangham 1986).

Wrangham's (1986) analysis of the differences between chimpanzee and bonobo social organization reveals the complex interplay between economic and reproductive factors in shaping community structure within each species. Moreover, Wrangham provides, at the end of his analysis, an *evolutionary* explanation, based on the supposition that bonobo social organization is derived from a chimpanzeelike precursor. Wrangham's explicitly evolutionary hypothesis is of note because much of the literature on the biology of sociality represents an application of principles of evolutionary theory to the analytic description purely of the structural and functional aspects of social systems—rather than an attempt to explain the history of the system, the traditional purview of evolutionary biology (see Eldredge [in press] for additional discussion on this point).

Wrangham (1986) succinctly describes the problem of explaining the great diversity in patterns of social organization among the four ape species: it is "attractive" because the ecology of the four species is so similar. Similarities include living in habitats that include "at least a small amount of tropical rain forest." Apes sleep in impermanent nests. Moreover, there are "no obvious differences in their vulnerability to predation, disease, or bad weather"; and all species "feed from rather discrete patches on easily digested food" (Wrangham 1986:352). Also included on his list of ecological similarities are common attributes of breeding behavior, such as the fact that the young travel with their mothers from birth, as well as the signal fact (discussed further in the next chapter) that mating in all four species occurs year-round.

Despite similarities in feeding behavior among the great apes, Wrangham believes that differences in foraging patterns account for among-species differences in great ape social organization. Among-species group size is not dependent on food density (i.e., bonobos always occur in far larger groups than do gorillas at any density of food resources). Thus Wrangham concludes that intensity of feeding competition—reflecting differences in distribution of different food types, and hence foraging style—is of material significance in determining group size and other aspects of social organization.

Wrangham (1986:365) interprets chimpanzee social behavior as dependent upon a "delicate socio-ecological" balance. Key to his explanation not only of chimp social organization but also of the differences among all great ape social systems is the hypothesis that "party" foraging reduces feeding efficiency (through competition). Thus, presuming that female reproduction is limited by food in-

take, females should spend most of their time alone (they do). (Like
Florida scrub jays and, indeed, many other vertebrate species, females typically emigrate from natal territories/communities, while
males tend to remain with the natal group.) Nonpregnant, sexually
active females without offspring likewise tolerate lowered feeding
efficiency in return apparently for the short-term benefits, including protection (from other females), food sharing, and grooming,
that are accorded sexually receptive females—though no long-term
male-female associations are ever formed in this promiscuous social
system.

Males, on the other hand, "tolerate" the reduced efficiency of
group foraging because (i.e., according to Wrangham's hypothesis)
male reproductive success depends to a great degree on male bonds,
which are nurtured in cooperative feeding behavior. Male allies
assist one another "against rivals during competition for status or
resources." Relative status mainly affects reproductive success: higher-
ranking males typically achieve greater numbers of copulations.
Males also cooperate in food-getting activities—though Wrangham
(1986:362) dismisses the idea that male bonding results from such
strictly economic cooperation. Thus association of males in foraging
parties acts simply to foster and maintain existing male bonds—
whose primary effects, or even purpose, Wrangham believes to be
to foster reproductive success. Interestingly, bonded males are not
usually close relatives.

Wrangham (1986:372 ff.) argues that the benefits of social bonding in the context of group existence in chimpanzees are primarily
"social" rather than "ecological." The benefit is primarily protection—whether it be for sexually active immigrant females traveling
with males; for females who travel with males to protect their infants from infanticidal attack; for maintenance or enhancement of
male social status, as when a bonded male helps fend off a challenge
from a third male; or for defense against male attack from outside
the community. It is to be noted that Wrangham's terminological
distinction here does not coincide with our own use of the terms
ecological and *social*. Wrangham (1986:372) means by *social benefits*
protection from attack by *conspecifics*, whereas *ecological benefits* refers
to protection from, or exploitation of, the nonconspecific biotic, as
well as the abiotic, elements of the environment.

We need not take issue with Wrangham's judgment that frankly
economic cooperative behavior among chimpanzees plays no role in
eliciting a need for male bonding (i.e., Wrangham sees the existence

of bonds as underlying such cooperation, and such bonds may well be imagined to be strengthened under such conditions of economic cooperation; but male-male bonds are *for* protection of status, which, in turn, is all about the maximization of reproductive success). Nor need we belabor the point that, however social versus ecological benefits from bonding are construed, it can certainly be maintained that protection from attack, whether by conspecifics (within-group females, within- or among-group males) or by nonconspecific predators, has a primary and direct *economic* effect. Only in the case of protection from infanticide can a reproductive gloss be applied in a totally straightforward manner.

Nevertheless, Wrangham's description and analysis of chimpanzee social organization constitutes an excellent example of the primary impact of economic behavior on social organization. In his scenario detailing how bonobo social organization might have been derived from a chimplike precursor, he must account not only for larger group sizes in bonobos vis-à-vis chimps, but also for the greatly reduced amount of within-group aggression. As Wrangham writes (1986:376), "Female bonobos spend more time with each other, have friendlier interactions, and have even evolved a new pattern of sexually stimulating greeting behavior. They spend more time with males than female chimpanzees do, are sexually receptive for longer, and have more elaborate sexual interactions. Males are less inclined to fight, groom each other less, and less given to travelling all over the community range than male chimpanzees are."

Wrangham attributes the derived condition of bonobos vis-à-vis chimpanzees to a change in their food resources, and hence foraging strategies. Bonobos (like gorillas, elements of whose social organization bonobo organization is said to resemble) use "terrestrial herbaceous vegetation" (THV) as a significant percentage of their diet—in addition to the tree fruits that form literally the bulk of the chimp diet. Wrangham simply suggests that such a mixed food-resource base reduces foraging competition, fostering greater social mixing—though he admits that a correlative reduction in male-male (reproductive) competition remains unexplained. It could simply mean that a de-emphasis on economic competition overall lessens male-male competition—suggesting further that patterns of both economic competition and cooperation between male chimpanzees may have greater strictly economic significance than Wrangham is inclined to admit. Because male-male bonds in chimps appear to have both economic and reproductive effects, and because exercise of such bonds in an economic context enhances the bond (with reproductive implications), it is reasonable to suppose the converse to be equally true: there would then be direct *economic* implications of an element of *reproductive* behavior. We pursue such themes in greater detail in our examination of the biological aspects of human social systems in the next chapter.

That there is a complex interplay between economic and reproductive behavior—and that the vector of causality is not strictly from the economic to the reproductive, but may as well embrace economic implications of reproductive behavior—seems quite apparent in great ape social organization. Certainly this is so in Wrangham's elegant evolutionary scenario deriving bonobo social organization through a simple change in mix of dietary items: an economic change effecting social change, especially elements pertaining to reproduction. Chimplike bonobo ancestors simply came to occupy a territory both rich in THV and free of gorillas. Description of the actual elements of chimp and bonobo social organization clearly reveals the close interplay between economic and reproductive activities of these creatures.

Social Systems Without Reproduction: Mixed Species Flocks

We have stressed in this chapter the overwhelming tendency of sociobiological theorists to see organismic reproductive behavior as central to the very structure and function, as well as to the history— in short, the very *definition*—of social systems. We have strived to counterbalance this view by stressing the coequal importance of organismic economic behavior with reproduction—and by arguing that social systems represent complex comminglings of economic and reproductive behaviors in ways not developed in nonsocial organisms.

To complete the picture, we briefly consider the obverse: existence of cross-genealogical aggregations that bear many of the hallmarks of social systems. Because reproduction does not occur within such systems, they exemplify the importance that economics per se has in the structure of any social system.

Mixed feeding flocks of birds provide the most striking example. Typically formed in the nonbreeding season, mixed flocks are often considered a phenomenon primarily of the tropics, but they in fact occur in all regions. Birds within such aggregations interact not only with conspecifics, but with birds of other species as well. The usual explanation for the formation of such mixed-species foraging flocks is protection (many eyes—with no possibility here of altruism through kin selection) and more efficient location of food resources (though, typically, not all avatars in such flocks exploit the same energy resources).

Such aggregations are often quite stable. Some such flocks form within breeding territories before migration (at which time, of course, the flocks dissolve as members of each species follow their own species-specific migratory timetable and direction). Others, such as winter feeding parties in high latitudes, remain stable for many months. The same downy woodpecker, red-breasted nuthatch, white-breasted nuthatch, black-capped chickadee, and tufted titmouse individual organisms can be seen roaming the same woodlot day after day. Larger, less closely knit aggregations occur at feeders, where birds not normally associated in foraging flocks may all be present at the same time (as when juncos and song sparrows join the chickadees et al.). Yet even in such patently artificial situations, the alarm call or sudden flight of any one bird is usually enough to send them all packing.

Most critically, birds continue to interact even when far from their breeding territories. Male juncos, for example, dominate females and juveniles and have, as well, a pecking order among themselves. Such appears to be the case for most, if not all, these winter, purely economic "microavatars." Even if it could be argued that maintenance of dominance patterns in the nonbreeding season simply keeps order until the breeding season returns (unlikely for juncos, which only form conspecific flocks after the breeding season), the dominance patterns of interaction among *non*conspecific organisms within mixed flocks must still be explained. A female downy woodpecker will shoo a male white-breasted nuthatch from a basket of suet, only to be shooed herself by a male downy woodpecker. If within-avatar (generally competitive) interaction still seems more intense than among-avatar interaction in such situations, the two seem to share the same fundamental aspect: both are about maximizing individual organismic access to energy resources.

These mixed-species (hence patently nongenealogical), purely economic associations are the exception that proves the rule about social organization. Such systems are not true societies, because reproduction is indeed one of the two prime aspects of true social organization. But, by factoring out reproduction, mixed flocks, mimicking elements of social organization, demonstrate the importance that purely economic considerations have in the structure of any social system.

Social Systems as Entities

In previous chapters we have discussed at length the ontological status of large-scale biotic systems: demes and avatars, species and ecosystems, etc. We characterized these as categories of historical entities of particular sorts. Arguing that entities such as these arise from the interactions of organisms putting their adaptations into play, we sorted these entities according to the kind of work they perform. Organisms do economic work, and that work leads to interactions of certain kinds—interactions that lead to the formation of local conspecific populations of interactors (avatars), themselves parts of cross-genealogical economic systems (local and more regional ecosystems). Likewise, reproductive behavior leads to the formation of aggregates of interbreeders (by definition, conspecifics) that we call demes. Demes are parts of species, which are in turn parts of larger genealogical entities, monophyletic taxa. Both kinds of entities, the economic and the genealogic, are hierarchically nested, with smaller-scale entities being parts of larger-scale entities.

We also concluded that, above the organism level, such biotic entities are of either one sort or the other. That is to say, organisms perform economic as well as reproductive functions. But *above* the organismic level, entities are *either* economic *or* reproductive (genealogical). However, there may be cases in nature—particularly in stationary animals (and, of course, plants)—in which the membership of demes and avatars overlaps to a degree, or even in which recognition of the two as separate entities is impossible and, more important, conceptually unnecessary (though we should never lose sight of the fact that both activities are present). We have conceded that the split between economic and reproductive entities at the local-population level is less than thoroughly complete. For one thing, we are dealing with *conspecifics* in avatars as well as in demes. Only above the avatar level (i.e., at the ecosystem level) do economic systems become wholly cross-genealogical, that is, nongenealogical. We maintain, however, that at this "local population of conspecifics" level, demes are indeed more commonly separable from avatars—at least more so than is generally supposed, as flocks of migrating birds reveal well: though conventional wisdom holds that "real," well-differentiated subspecies must be allopatric, members of different subspecies of short-billed dowitchers rub shoulders on intertidal mudflats regularly during migration. There is no problem with these taxa mingling in a purely economic context; when reproducing, these subspecies occupy distinctly separate regions of northern North America.

In the context of the previous discussions of the status of demes and avatars, we now ask, what manner of entities are social systems? Like avatars and demes, social systems (and marine invertebrate colonies) arise from the interactions of component organisms. But, as we have endeavored to establish in this chapter, instead of the entity arising as a consequence of one class of activity (i.e., economic versus reproductive) or the other, we find in social systems that it is the complex interaction *between* economic and reproductive activities that leads to social cohesion. Whether it be in a metazoan or metaphyte body, a marine colonial invertebrate, or a eusocial insect system, there is a pattern of assignment of tasks (division of labor), with the primary division between economic and reproductive activities (with bridging activities, as in "child care" in eusocial insects). Such systems contain no "stand-alone" organismic component; each organism (or polyp or zooid) contributes in some particular wise to the total effort of the system.

In vertebrates, the nature of the division of labor is less clear-cut, and, in particular, there is a far greater tendency for any given component organism to contribute *both* reproductively and economically. Yet no matter *how* these interactions are organized, the point is that economics and reproduction are interwoven in social systems, so much so that the whole "point" of social systems seems to be the integration of these two activities—activities that otherwise tend to create large-scale biotic systems that are either economic or genealogical in nature.

In using the word *point*, we are, of course, asking why it would occur in evolution that these two classes of activities are integrated in such fashion(s)—and, perhaps, why it has occurred in some taxa but not in others (though such questions are problematic in evolutionary discourse in general—i.e., asking why something has *not* evolved in some groups). Beyond the (rather vague) suggestion that interlocking economic and reproductive adaptations and localizing them into a single spatiotemporally bounded system could reasonably be considered a step to greater "efficiency," we have no further answers to the evolutionary "why" of social systems. We have been content, in this book, to characterize the basic elements and natures of large-scale biotic systems and to locate social systems in the scheme of things. We see such systems arising, on a small-scale, case-by-case basis as well as on a phylogenetic basis, from the interactions among and between the economic and reproductive activities of organisms.

There is (to bring the analysis of social systems to the same degree of completeness as that of the elements of the genealogical and economic hierarchies) no difficulty in seeing social systems as spatiotemporally bounded, (historical) entities. They have beginnings, histories, and ends. This is as clear for beehives as it is for hominid bands or any other specifiable element of human social organization. And, because social systems represent a fusion of economic and reproductive elements, it follows that the actual work of any social system is both economic and reproductive. A eusocial beehive does obvious economic work within its surroundings; it is, in short, an avatar within a local ecosystem. But it is also reproductive—and it is so *on its own level,* that is, beyond the obvious fact that more individuals are regularly being produced within the hive. When queens leave to establish new colonies, the social system is in effect reproducing. Wilson (1985) has described huge "supersystems" of ants, in which many colonies are living in close, interactive proximity—an element of hierarchial organization as well to eusocial insect organization. Because eusocial insect colonies are both avatars and demes, and because such entities interact economically and reproduce as entities, no conceptual difficulty impedes seeing them subject to "avatar/deme" selection, that is, with components analogous to *both* natural and sexual selection at the organismic level.

Vertebrates pose a somewhat different picture. When young male beavers leave a pond for bluer waters, they are founding (when they find a female, that is) new colonies. And, in so doing, these new beaver pairs are founding a socioeconomic/reproductive entity— one that is definitely a component of the local ecosystem (is an avatar) and is, as well, a reproductive entity. But a beaver deme, because it involves regular out-breeding (and most vertebrate social systems, of course, do) is de facto larger than any single colony (avatar). Likewise, Carneiro (e.g., 1970) has described regular fissioning within human tribes in the Amazon Basin, where villages will fragment in response to growing population size in relation to available food resources: vertebrate social systems can and do "make more of themselves." There is, for vertebrates, a less exact matching of avatar and deme rolled up into a single, neat social package than is the case in eusocial insects. The same holds, of course, for human societies. Yet, it must be stressed, the reproductive and economic needs of individual organisms are regularly and largely met within the context of local social systems or through interaction (for reproduction) with immediately adjacent systems.

Nor are vertebrate social systems hierarchically structured precisely like insect systems—which should not be surprising, given the fundamental manner in which the two kinds of systems differ in the interrelations between economic and reproductive organismic activities. True, an aerial photograph of the Nile Delta shows a regular array of villages, which are arranged (as are towns and villages the world over) into a hierarchical array of political entities; this is an element of (primarily, perhaps exclusively) economic hierarchical structuring. And the reproductive interactions among adjacent social systems in virtually all social vertebrates introduce an asymmetrical element of hierarchical organization into their social systems. But these examples are *either* economic *or* reproductive, and they do not offer the fusion of both in the manner of the vast supercolonies of ants (Wilson 1985), in which the whole emerges as a grand version of its interacting component parts.

And, though all humans live within a sociopolitical hierarchical structure that is precisely geographic (i.e., exactly like the economic systems that all organisms find themselves in, but one, we add, that does not match the existing economic systems in nature), that by no means exhausts the kinds of hierarchical systems in which humans find themselves. As we have already seen, these systems are almost entirely economic—as when someone is a member of a department within a firm that is, in turn, part of a conglomerate. The systems of hierarchical structures in which humans in the larger and more complex social arrays invariably find themselves cut across one another in a kaleidoscopic, if not utterly bewildering, fashion. Such an interlocking system of (primarily economic) hierarchical arrays is unknown in any other form of life.

The point of this chapter has been that social systems represent fusions of the economic and reproductive behaviors of organisms. Organisms are both economic and reproductive entities; their behaviors are connected but lead, in biotic associations, to the simultaneous formation of two different sorts of systems: economic and reproductive (genealogical). Social systems retard this split. Presumably in the interest of efficiency, the economic and reproductive needs of organisms are brought closer together, forming cohesive social systems that are both economic and reproductive in character. The fusion is effected in very different ways: metazoan bodies, colonies, and eusocial insects are basically similar in that there is a fundamental division of labor between replicative (reproductive) and economic functions, and, further, among various sorts of economic tasks. The mode for such division is assignment of individual members (cells, polyps, individual organisms, as appropriate) to specific tasks. Vertebrates, including humans, retain greater flexibility in assignment of tasks—great redundancy in economic and reproductive roles, which specifies a rather different set of interactions in social organization.

The sociobiological paradigm is derived directly from the ultra-Darwinian contention that life is organized around an eternal competition to spread alternate programs for its (pro)creation. The supposition is derived from the correct observation that differential economic success leads, ceteris paribus, to differential reproductive success. There is, however, no reason to conclude thereby that economic aspects of life exist only to support the reproductive. It might, perhaps more compellingly, be argued that life is a self-organized system of organic molecules capable of extracting energy from the surrounding environment—a temporary dissipative structure. Because (by the nature of their materials, and ultimately, the constraints described by the second law of thermodynamics) such structures *are* temporary, only those that also could make more of themselves managed to survive—as genealogical systems.

It seems a more balanced description of what living systems *are* to say simply that they typically involve economic *and* reproductive aspects. It seems, further, that to understand the history of systems, how they came to be, entails knowing how they are arranged and how they function. Even the cursory description and analysis of colonies and various social systems that we have been able to provide in this chapter demonstrates that social systems arise as fusions of reproductive and economic behaviors of organisms. Unlike avatars and demes, and ecosystems and species, social systems remain as parts of both the economic and genealogical hierarchies.

CHAPTER SEVEN
•••••••••
Human Sociality

So far, we have been developing an evolutionary, or, more generally, a biological framework for the study of social systems that we find more adequate than the more narrow ultra-Darwinian approaches still fashionable in many places. We want, in conclusion, to apply this model (with some necessary complexification) to the case of human societies. Let us first recall its major features.

Our model differs from previous approaches in two respects (we have said this often, but should perhaps say it yet once more at the head of this, our final chapter). First, rather than being committed to one unique level of biological entities, be they genes, organisms, or ecosystems, we have been arguing that the reproductive and economic activities of organisms produce hierarchically arrayed entities that must be taken into account in various ways in an adequate evolutionary biology. To take selection as targeted to only one level, for example, that of organisms, may be tempting—it is at that level that selection is most easily observable, whether in the wild or in the laboratory. And that bone of contention, species selection, may be difficult to demonstrate. Yet openness to the existence of biological entities at more than one unique level is necessary, in our view, if we are to take account of the complex variety of phenomena that biologists in many disciplines observe and attempt, with some success, to explain. Thus we share with a number of others the belief that, within appropriate contexts, it is wise to take a hierarchical view of biological phenomena.

Second, and more especially (since we have not [yet] many comrades in this aspect of our venture), we have been stressing the existence of two separate, if interacting, hierarchies: the genealogical and the economic (or ecological). Even those others who make the same distinction may wish to make it in a different fashion (Mishler, pers. comm., 1988). Our distinction, however, if out of line with some others, should by now be clear: within a view of evolution that is basically Darwinian and therefore "adaptationist," we wish to distinguish two kinds of adaptations that characterize living things in general and animals in particular: adaptations bearing directly on reproduction and adaptations bearing primarily on means of making a living in a given environment. Thus Darwin's distinction between sexual selection and natural selection stands at the very center of our model. Even this division will be complicated in the human case, as we shall see, by the separation in human life between sex and reproduction. But complicating the division does not undermine it. Leaving that complication aside for now, however, what we have been arguing for social systems in general is the thesis that social life, whether in its insect or mammalian form, has arisen from the reintegration, in a variety of ways, of that basic separation. The reunion of the reproductive and economic activities of organisms permits the emergence of social systems: that is our basic theme.

As we said, we want now, in conclusion, to apply this model to the human case. That is no easy task. If, as we noticed earlier, recurrent conceptual tensions exist at the base of Darwinism, so much the more evident are these or similar contrarieties in attempts to account, whether philosophically, anthropologically, or sociologically, for the nature and origin of human social systems. Theories of human society may be more or less atomistic, more or less focused on totalities. They may be more or less deterministic, permitting less or more to the power of individual agents. In general, however, it seems that emergent social science has sought to imitate its vision of "harder" sciences. Science is thought to be impersonal; if persons are to be described, and their actions explained, scientifically, they must also be treated impersonally—that is, overlooked or negated in a scrupulously objectivist account. This aim is sometimes implemented through a deterministic account of individual behavior, as in the behaviorism of Watson or Skinner, sometimes by a global model that overrides the very existence of persons, as in structuralist anthropology. Overall, it seems to be the effort to achieve pure objectivity in the face of our strange "subjective" interest in our own concerns that has driven much of social science to formulate its theories.

Such pure objectivity, however, is one of the idols of our particular society. There is nothing viciously anthropomorphic in human beings' concern with humanity. The recognition that persons exist as centers of action is a necessary component in any adequate account of their behavior, in groups as well as singly (not that they are ever wholly "single"!). For that matter, every scientific investigation in every field arises within the limits set by some area of concern, some vision of the way things are. The image of nature as a machine helped to shape classical physics and, as we have seen, still drives some versions of Darwinian thought. Our more ecological and hierarchical model is equally an image. It expresses fundamental beliefs, which, of course, we believe to be well grounded in biological reality as a number of reliable disciplines have come to read it. But these are beliefs, if reasonable ones, not wholly indubitable assertions of pure objective "fact."

Fortunately for our readers, this is neither the time nor the place to weigh such global problems in the philosophy of science. What concerns us here is simply the recognition that our approach to the question of human sociality should aim to transcend the would-be objectivism of many accounts, not by falling into the trap of "subjectivism," but by moving from within the recognition of our phylogeny to ask what, if anything, is human in human social systems and how they are best to be characterized as one among many mammalian and one among many primate versions of sociality.

Needless to say, we are not seeking one "essential" characteristic of every human being, or even of every human society, to mark us off, simply and forever, from our cousins among primates or mammals or vertebrates. The organisms that count as members of a given species can not be characterized by any *one* property.[1] Beavers are not "simply" dam builders, any more than human beings are "simply" human-language users. But because we have noticed some striking differences between insect social systems and those of vertebrates, we want to ask what, if any, differences single out our style of social life from that of other vertebrates.

Social Science in Evolutionary Perspective

That social systems, including human social systems, have come into being as the result of evolutionary processes should not need stating. Societies, even of our peculiarly culture-dependent variety, can be counted among the natural existents on this planet. Like genes, organisms, species, or ecosystems, human societies or, better, the conditions permitting the existence of human societies, have arisen as the product of what Jacob (1970) calls evolutionary tinkering. Many believe, therefore (although in our view it is a questionable "therefore"), that the classical theory of natural selection in its linear, ultra-Darwinian form accounts adequately for all human activity. The recently founded Society for Evolution and Behavior appears to be proceeding on this assumption. We should remember, however, that the principle of natural selection, while itself a powerful heuristic in many different biological circumstances, is in itself always in need of particular empirical content (Brandon 1981). As Washburn (1978) has convincingly argued, such data are often lacking in global attempts to extend "selection" from nature to the phenomena within nature that we call culture. He writes:

> The problems of evolutionary theory and social anthropology may be illustrated by the confusions of some of the leaders in the field of evolutionary biology. Alexander (1975) concludes that society is based on lies. Hamilton (1975) thinks infusions of pastoral genes are necessary to keep creativity. Wilson (1975) thinks neurobiology is necessary for ethics. In addition to technical disagreements, especially over group selection, these authors sound like a mixture of Herbert Spencer and science fiction.
>
> Various reasons lie behind the confusions of evolutionary theory when it is applied to human beings. Perhaps the most fundamental is the belief that if the theory is correct the application of the theory is easy and routine. But some general theories—the notion of selection leading to adaptation, for one—are only guides for research. If human locomotion is the area of interest, then the structure of the pelvis and the lower limbs must be interpreted in terms of the way they function. This adaptation is unique to humans, and there is no way that this form can be predicted from the theory. Nor is the form resulting from the evolutionary process necessarily efficient or ideal. It is just what, as a part of a complex process, succeeded in competition with others. . . .
>
> The application of the general theory of evolution (selection, adaptation, inclusive fitness) requires that the facts used in fitting the

theory itself only suggests where to look for facts and which ones may
be relevant. This is the fundamental weakness of the authors I have
cited. They know theory, but they know very little about human
behavior or the less known facts of evolution. The power of the
theory gives them confidence that their personal opinions are correct
and should be adopted. (Washburn 1978:63–64)

Of course we have our personal opinions, too. But we hope that
our proposed expansion and complexification of Darwinian theory
allows us a more adequate account of the nature of social systems in
general and some insight into the character of human social systems
in particular. As the preceding chapters have recurrently stressed,
we find evolutionary orthodoxy still too narrow in its premises and
methods to allow an adequate account of the complex and varied
phenomena that the history of life has produced including ourselves. We have been trying to forge a conceptual mold that will
permit a richer and more appropriate account of social systems,
including, among others, human social systems.

Are We Different?

Given our general position about the nature of social systems, we
must ask whether there is, as many have claimed and many have
also denied, something unique about human societies. We exhibit,
of course, one variant on the theme of mammalian societies, one
variant on the theme of primate societies. Do we differ from these
kin of ours in any significant way? We will look in two directions for
an answer to this first question: first, by pointing to some very
obvious empirical observations; second, by looking in particular at
recent work on primate communication. Having assured ourselves
that as animals within nature we have nevertheless developed a life-
style sufficiently distinctive to be worth characterizing, we will then
try to say something by no means definitive or dogmatic, but still
something, about the peculiar nature of this human life-style.

One more brief warning is in order, however, before we proceed
with this project. We have been attempting, in the preceding chapters, to fill in the biological *context* for the existence of animal societies, including our own. That does not mean that we have been
constructing a scenario for the actual evolution of such societies,
and here, in particular, not of human social evolution. The fact of
evolution is indeed a basic presupposition for our endeavor: that we
have come into existence through an evolutionary process probably

largely Darwinian in character is assumed in our reflection on the peculiar version of social life that marks our life-style. With respect to organic evolution, we are asking what conditions have probably made possible the emergence of human sociality. Whether, and at what date, we descended from Lucy or some other African ancestress is not a question we are concerned to answer here. Again, it is the evolutionary context of our sociality we start from, and it is the character of that sociality we want, tentatively and perhaps somewhat timorously, to describe.

With that warning in mind, let us proceed to ask whether human society is distinguishable from the social systems of other animals, and if so, how. So we must ask whether, as social beings, we do indeed differ in any interesting way from our nearer or farther kin. Hume in his *Treatise of Human Nature* appended a section on "reason in animals" and two on various "passions" (i.e., emotions) in animals to his arguments on human reasoning and human emotion (Hume [1739] 1955, bk. 1, pt. 39, sec. 16; bk. 2, pt. 1, sec. 12: bk. 2, pt. 3, sec. 12). His theory, he boasted, was superior to others because it needed only one hypothesis to account for all animal behavior. That sense of continuity is of course reinforced by our acceptance of evolution and thus of our awareness of kinship with other animals. Thus students of animal behavior often claim to have plumbed the depths of human nature by looking at their subjects, be they herring gulls, lizards, or what you will (Tinbergen 1960). And we can see what they mean: curiosity, exploration, pride, shame, jealousy all seem to be as observable in other animals as in ourselves. Perhaps the sense of human uniqueness was tied to the notion that something other than animal, some immortal soul or mind, had been inserted into the human neonate or embryo or egg. To most, if not all, biologists, however, such a notion seems improbable to the point of absurdity.

Moreover, every time that people have claimed some unique character for human life, it has broken down. Descartes thought the linguistic distinction entirely obvious; if it holds (we will return to that question below), it certainly does not do so in the simple way he thought (Descartes 1973b). There was tool using, but the Gombe chimpanzees and others have clearly eliminated that (Goodall 1986). There was recognition of mirror images, for which Gallup (1970) found clear experimental refutation. Recently, an account of human social organization has been based on the separation, in our case, between sex and reproduction (Wilson 1980; Fisher, 1982),

but that, too, as well as frontal copulation, turns out to be characteristic also of our very close relatives the pygmy chimps (Savage-
Rumbaugh et al. 1978). And, as we noticed in chapter 6, year-round
receptivity to sex is found in other great ape females as well.

If, then, as individual animals we are not special, why should our
societies be special? Granted, vertebrate societies—and in particular,
mammalian societies—are, for the most part, differently organized
from those of insects. But we are mammals—and not naked mole
rats. Why shouldn't we accept a generalized account of mammalian
social systems and count ourselves among them? That is what animal behaviorists often claim they do. We are just one species among
others; nothing special about us.

Yet there do seem to be differences in human social behavior as
compared to other primates or mammals that suggest a fundamental distinction between our social systems and theirs. Peter Wilson,
for example, has founded an ingenious account of human sociality
(of which we will have more to say below) on the distinction between
sex and reproduction, as has Helen Fisher (Wilson 1980; Fisher
1982; cf. Harris 1989). True, they ignore the evidence for such a
separation in the other great apes, especially the pygmy chimp. But
perhaps *Pan paniscus* has not taken this separation as far as we have;
or perhaps in humans the suppression of estrus has combined with
excessive cerebralization, more constant upright posture, and so on,
to produce a situation, both reproductive and economic, that demands more exotic social organization than that of pygmy chimp
societies.

Whatever the case in this respect, on the surface our social systems do seem to differ rather plainly from those of other primates.
Most obvious is the great variety of social structures characteristic of
human cultures. Perhaps there are some universals of human society, but it is the differences that are most striking. No other mammals, so far as we know, display such diversity of social organization
within a given species. We are, like members of many other species,
"naturally" social animals: we need one another. But the devices by
which we organize and control our mutual need varies amazingly
from group to group. To an extraordinary degree we have to make
ourselves, if always within a social construction that at the same time
makes us. As Pico della Mirandola makes God tell Adam, "We have
given you, Oh Adam, no visage proper to yourself, nor any endow-
ment properly your own, in order that whatever place, whatever

form, whatever gifts you may, with premeditation, select, these same you may have and possess through your own judgment and decision" (Pico della Mirandola 1956:7). In comparison with other primate species, moreover, we appear to have developed a more complex social system. Gibbons, for instance, form stable family bonds but have no larger social groupings. Chimpanzees, on the other hand, form larger groups but lack unique family structure (Wilson 1980). Only humans seem to combine both kinds of organization: family and society. Even more striking is our apparent need for the coercive structure of the state; at least in many cultures, we have to devise complicated and arbitrary sanctions in order to control our own violence toward one another. We are indeed social animals, but each of us also retains an asocial, even antisocial, tendency that demands arbitrary control. Thus we are not only social but also *political* animals. Quite apart from our own natural inclination to be concerned with our own case, therefore, all these very obvious phenomena suggest that some special explanation of the structure and functioning of human social organization is required.

The phenomenon that has seemed, through the centuries, most clearly indicative of our uniqueness as social beings has, of course, been language, or our kind of language. That seemingly obvious distinction, however, has become controversial, and we need to take a stand on this issue before we proceed further. In terms of molecular biology, our nearest relatives, the chimpanzees, are very close kin indeed, and even before the advent of molecular techniques experimenters had been trying to demonstrate that these our cousins, in addition to resembling us uncannily in other ways, can speak. As Duane Rumbaugh summarized the ape language work a few years ago:

> They (sc:apes) can learn words spontaneously, efficiently, and they can use them referentially for things not present . . . ; they can learn words from one another . . . ; they learn to use their words to coordinate their joint activities and to tell one another things not known . . . ; they can learn rules for ordering their words . . . ; they do make comments . . . ; they announce their intended actions . . . ; they are spontaneous and not necessarily subject to imitation in their signs and not necessarily poor turn-takers in conversation . . . ; and they are communicative and they know what they are about. (Rumbaugh 1985:41–42)

He concludes his summary by declaring: "Apes are in the language domain, a behavioral domain that is a continuum—not a dichotomy" (Rumbaugh 1985:42) But since language is clearly the major carrier of our social systems, such a statement suggests that ape society too is really just like ours, different only in degree, not kind. The work of Sue Savage-Rumbaugh with those even more language-friendly animals, the pygmy chimps (*Pan paniscus*), supports this view; she believes that if she could but study pygmy chimps and pygmies in their neighboring native habitats, she would find only a slight difference in the degree of enculturation—and thus the myth of the uniqueness of human culture would be refuted once for all (Savage-Rumbaugh, Rumbaugh and McDonald 1985; Savage-Rumbaugh, oral. comm, 1989). If that is so, what are we fussing about? Who are we to make such a fuss?

Granted, there may still be some linguists, notably of the Chomskyan school, who insist on the uniqueness of human language in its "deep structure"—something really different written in our genes. There is always some syntactical feature of our language—modalities, double articulation, or what you will—that seems to evade the relatively simple-minded, or single-minded, conversations of our simian friends. But then there is always some experimenter who will show they can after all do that—or that—or that. The debate seems endless.

The trouble is that all the efforts to establish human uniqueness, particularly with respect to language use, have been addressed to the question of communication. As communicators we are indeed just somewhat "cleverer" than other animals; there is no difference between us except one of degree. This holds as much for communicative as for other forms of behavior. However, as Sahlins has argued in a much-abused book, *The Use and Abuse of Biology* (Sahlins 1976:61), what is unique about human language is not its *function of communication,* but its *structure of signification;* its aim, fundamentally, is not so much to *convey information* as to *generate meaning.* Sahlins writes:

As communication, language is not distinguishable from the class of animal signaling, it only adds (quantitatively) to the capacity to signal. What is signaled is information—which may be measured, as in classic theory, by the practical alteration in the behavior of the recipient from some otherwise probable course of action. . . . This functional view of language, which incidentally is exactly Malinowski's, is particularly appropriate to a biological standpoint, for by it human speech is automatically subsumed in the adaptive action of responding to the natural or given world. What is lost by it is the creative action of constructing a human world: that is, by the sedimentation of meaningful values on "objective" differences according to local schemes of significance. So far as its concept or meaning is concerned, a word is not simply referable to external stimuli but first of all to its place in the system of language and culture, in brief to its *own* environment of related words. By its contrast with these is constructed its own valuation of the object, and the totality of such valuations is a cultural constitution of "reality."

What is here at stake is the understanding that each human group orders the objectivity of its experience, including the biological "fact" of relatedness, and so makes of human perception and social organization a historic conception. (Sahlins 1976:61–62)

Notice that human experience here becomes "historic" in a way in which the processes of evolution, the age-old story of evolution, are not so. Human history is, so to speak, history doubled. If evolutionary tinkering has produced the myriad life styles that have been, and a comparative few of which still are, characteristic of life on this planet, human construction has produced, within the constraints of that biological reality, a host of new "realities," symbol systems within one of which each member of our species achieves, or fails to achieve, responsible personhood. We have been freed, within limits, from the tyranny of the gene, only, it seems, to be subjected to the tyranny of a culture. Yet it is only within the limits of such a further, artifactual control that we can achieve characteristically human lives. Note: this is not relativism or subjectivism but "perspectivism," based on the recognition that each culture constructs its own reality *within* nature, using natural materials (including our brains and hands and voices) to make the "second nature," the cultural home, within which human beings are enabled to develop into the responsible agents they have, other things being equal (as often, alas, they are not), the potentiality to become.

What is important to recognize at this juncture, however, is not so much the necessary priority of culture to human agency as its priority to the very objectivity of objects. Sahlins continues, in the same passage:

Human communication is not a simple stimulus-response syndrome, bound . . . to represent the material exigencies of survival. For the objectivity of objects is itself a cultural determination, generated by the assignment of a symbolic significance to certain "real" differences even as others are ignored. On the basis of this segmentation or *découpage*, the "real" is systematically constituted, that is, in a given cultural mode. (Sahlins 1976:62)

A passage Sahlins cites from an obscure paper by Cassirer is worth quoting here. Cassirer writes:

> An "objective" representation . . . is not the point of departure for the formation of language but the end to which this process conducts; it is not its *terminus a quo* but its *terminus ad quem*. Language does not enter into a world of objective perceptions already achieved in advance, simply to add to given individual objects signs that would be purely exterior and arbitrary. It is itself a mediator in the formation of objects; it is in one sense the mediator *par excellence*, the most important and valuable instrument for the conquest and construction of a true world of objects. (Cassirer 1933; cited in Sahlins 1976:62–163)

This is an extension and transformation, within the perspective of modern evolutionary biology, of Kant's fundamental insight: that it is our conceptual activity that makes experience objective. Or as he put it, classically, in the Transcendental Analytic: The grounds of the possibility of experience itself are at one and the same time the grounds of the possibility of the objects of experience (Kant [1787] 1968, vol. III, B 197).[2] This should be obvious for such sophisticated human activities as the arts and sciences. Whatever section of the natural world we enter—whether it be the world of the honeybee's dance, the world of subatomic particles, the world of plate tectonics—we enter it through some human subculture that enables us to project ourselves into that foreign domain. Even in our more mundane concerns, such as the search for mates or our preferences in food, clothing, and shelter, the constructions and choices of a particular human society constrain and positively shape our behavior. And language is the most massive and obvious vehicle for such constructions and choices. Thus the primary task of human language is world building, object forming, rather than the relatively simple, relatively unequivocal task of giving someone some bit of information, such as "food source in that direction" or "this individual ready for mating now."

We will have more to say about enculturation and human society below; here it is a question of replying to the "chimps-do-it-too" theme. Let us state our response baldly; we will expound a little more fully (though still sketchily enough) below. For now we are just endorsing the distinction made by Sahlins between linguistic communication and language-based world building. However many vocables Sherman, Austin, Sara, or Kanzo learn, however much they enjoy using them, however (relatively) complex the situations in which they use them, however much they "know"' what they are doing, they do not appear, in our view, to be entering into the construction of a world of significances in the way that human infants-times-toddlers plainly do. They give a lot of signals to the experimenters and to one another. And to that extent they enter to a degree into *our* world; indeed, they adjust suprisingly well to this new environment. But it is not only a world they never made—so is the baby's—it is also a world no one in their species made, and they do not participate in its remaking as the beginning human language user, in cooperation with her caretaker(s), appears to do. Chimp speakers acquire new tools, of course, and so do babies. But a child becoming a language user enters a whole new world, a totality of symbols and therewith symbol-imbued behaviors that, for all our biological kinship, remains unique. We live *in* language. It is that artifactual home, grounded in but not determined by our biological needs, that distinguishes us most obviously from other primates, other mammals, other animals. Again: if, within limits, we are freed from the control of our genetic program, each of us, and each society, is subject to the control of a second, invented program within which we live, move and have our being—and which, in turn, by such living, we constitute and transform.

Further, language, although it is the most massive and evident expression and bearer of this oddity of our life-style, is nevertheless not its most fundamental ground. Human language shows and mediates the process or activity of *symbolic activity* or *symboling* through which our societies are constituted and constitute themselves. We will try to explicate and illustrate this theme below. Before we do that, however, let us first provide support for our position by noting some of the biological pecularities of the human animal that seem to underlie or support our style of social life and then try to consol-idate the contrast we have been suggesting by putting it in terms of a particular version of what is sometimes called philosophical an-thropology. Granted, what is being characterized in such an account is not as such human social life, but any individual human life. But as we shall see, the two—a human society and a human person—are indeed inseparable. If we can characterize briefly the way in which our life-style differs from those of other animals, we will have better access to the difference in human social systems that we are trying to single out.

Biological Foundations of Human Sociality

If species are historical entities, as we take them to be, each one, patently, is unique. At the same time, since they are related as family members, they may also have much in common. In general, we find specialized exemplifications of common themes: as, for example, variants in plumage, beak size and shape, leg and claw length and size, and general body size and pattern in birds (Grant 1986), or in behavior, differences in courtship rituals, nesting behavior, or parental care (for the last, see Winkler and Wilkinson 1988). Given avian generalities, there are species-specific characteristics—not only an SRMS, but economic characteristics as well—by which one interbreeding group can be distinguished from others (as some of Darwin's finches, for example, feed on different vegetation from others in what looks at first like the "same" habitat [Grant 1986]). In our case, we can tell *Homo sapiens* from our primate cousins in two ways: by a few specializations, analogous to the specialized beak size, say, of a given species of Darwin's finch (Grant 1986); and by generalization, the very contrary of specialization. Let us recall, in a summary fashion, the items on both sides of this ledger.

Anatomically, our signal specializations, all of which provide necessary conditions for the development of life in a human world, include the opposable thumb; the overall anatomy of the hand; bipedalism and the plantigrade foot; the larynx, specialized for speech as that of the chimpanzee is not; and, of course, cerebralization—both the change in proportions of areas in the brain and (apparently) the occurrence of increased asymmetry in brain localization. Our perceptual systems correspond fairly closely to those of chimpanzees: poor noses; excellent ears and eyes, the latter especially gifted in color vision; sensitive hands (Passingham 1982). Of course, as in the case of any species, our several specializations have evolved together as they have served to enhance our species-specific manner of coping with the contingencies of our environment. Cerebral language areas, if not unique to humans, have certainly become much more highly developed in our case than in others; and plainly, the evolution of the larynx (which in the human embryo is still chimpanzeelike) has closely coevolved with the speech areas of the brain.

Human specializations, then, in large part support the evolution of human language and the forms of social life developed and expressed by its means. Yet, as has often been noted, most conspicuously in recent English language literature, in Gould's *Ontogeny and Phylogeny* (Gould 1977), it is retrogression from specialization that most conspicuously marks the human form(s) of life. This is the paradox in human evolution: our most conspicuous specialization is the loss of specialization.

In many ways we appear to have evaded the specializations of other species by remaining, in effect, ape babies all our lives. Gould recounts past efforts to support, or in some cases to refute, such a view, beginning with Bolk's fetalization theory. It is the general pattern of retardation in our development, he argues, rather than some itemizable list of paedomorphic features, that has permitted our evolution as the odd animals we are. So, although Bolk practiced that kind of list-backed rhetoric and although he spoke out of what now seems a quaint, internalist evolutionary theory, his basic insight was correct. At some time some descendants of the ancestors we share with chimpanzees failed to grow up and so preserved the wider potentialities of infancy that permitted the initiation of the human condition. Why this happened is matter for mere speculation. Possibly, as Gould suggests (cf. Fisher 1982), the development of upright posture made giving birth to really full-term neonates too difficult, and so rather half-baked young became more common in populations of walkers.

Again, just what natural selection accomplished and just when, it is unnecessary to unravel—if we could. The point is simply that, as with any other species, the splitting of the lineage that led to *Homo sapiens* from the lineage that led to *Pan* initiated a series of events that ultimately led to the complex life-style we recognize as ours: a life-style grounded in new anatomical, physiological, and behavioral features, in this case features in large part associated with a new, retarded rhythm in development. In most respects such an alteration simply extended quantitatively a tendency already begun in other cases. Gould emphasizes that developmental retardation is already characteristic of primates; we have taken that change a step further. But that further fateful step in the rhythm of temporal development has permitted the spatial extension of our achievements, or predations, if you will, to the whole surface of the globe and a little way beyond.

Our oddity as animals, in particular in the condition of our neonates, was analyzed some thirty years ago in the work of the Swiss zoologist Adolph Portmann and his coworkers, to which Gould refers in passing (Portmann 1956; cf. Grene 1974).[3] Portmann made a study of comparative development in vertebrates and also in particular of the developmental pattern of human individuals as compared with chimpanzees. Postnatal development occurs in animals in two ways, generally distinguished as altricial or precocial: altricial species are those whose young are born relatively immature and remain for a period "in the nest" before venturing independently into the world (in German they are *Nestthocker*, literally, nest-huggers); precocial species are those whose young are relatively free-moving and even free-feeding at birth. Among birds the more "primitive" species are precocial, and the more highly developed produce helpless and sightless young needing a protracted course of parental care to prepare them for the adventure of living. Among mammals, evolution has gone the other way. The young of more "primitive" (plesiomorphic) species are nest-dependent, while the newborn offspring of "higher" mammals are ready with their first breath to disport themselves on their own four feet. One has only to compare a newborn kitten with a newborn lamb to see the difference.

Associated with this striking difference in mobility are two other important contrasts: the young of altricial species are usually born in large litters and have relatively undeveloped sense organs at birth, while precocial young are fewer in number per pair of parents but as wide-awake to their surroundings at birth as they will be at maturity. A comparison of the ontogeny of the two types illustrates the evolutionary process involved in the last of these contrasts. In the embryos of altricial species, the eyelids form early and then close, and birth occurs before they reopen. In precocial species, the stage of lid closing, now meaningless in the life of the organism, is retained, but birth is postponed till later, when the eyes have once more opened.

Now, where do we fit into this classification? Nowhere, it seems. We are born helpless but with our eyes and ears wide open (the ears, it now appears, even before birth). Even though a baby does not focus on objects for the first few weeks, (s)he does respond to colors very soon indeed and to sounds as well. (S)he takes from the first moment a powerful sight of notice. In other words, we have retained the sensory characteristics of the "higher" mammals (as well as the small number in our litters!), but have reverted in other respects to an earlier pattern in our infantile way of life. We have become *secondarily altricial*. Thus the evolution of mammals as a whole shows a change of course: from altricial to precocial and then back again to a new kind of altricial state: a state of cranial precocity and postcranial "altriciality."

That this is indeed the case is strikingly confirmed in comparative studies of postembryonic development by Portmann's associates at Basel. In altricial animals, as one might expect from the condition of their sense organs at birth, the brain must develop to a size from eight to ten times its size at birth; for precocial species this figure is on the average 1.5 to 2.5—somewhere around 2. This holds even for our nearest kin, the anthropoid apes; they are still precocial— for even though the baby clings more closely to the mother than do other precocial animals, it does so, so to speak, by its own volition: it is free-moving. Moreover, their brain growth is still "normal," approximately doubling from birth to maturity. With us, however, this situation has been sharply altered: our brains have to develop from birth to maturity to *four* times the natal size—a clear reversal of this evolutionary trend. Moreover, the proportions of a human infant are much further from those of the adult than are the proportions of a newborn ape from the mature individual. At the same time, the human infant is much heavier at birth than any newborn anthropoid—almost twice as heavy. On all these counts, our pattern of development is unique among mammals.

What does all this mean? Let us look at the comparison a little differently. Instead of comparing the newborn young of each species, let us ask: when does a human being reach, in behavior and in physical proportions, a stage of development comparable to that of the newborn young of other highly evolved mammals? It is usually at about a year that the child takes his (her) first steps and so assumes the upright posture characteristic of our kind, and at this age also his (her) build has come somewhat closer in its proportions to the eventual adult shape. Also at about twelve months, Portmann alleges, the child begins to act like a person: to speak and to perform what we recognize as genuine, reflective actions. Taking the pattern of precocial development as our model, therefore, we see that the human individual is born prematurely and achieves the status of an ordinary precocial neonate at about twelve months of age. This hypothesis is further confirmed by a comparison of the rate of growth of the human baby and the young of closely related species. The apes show a steady development from birth to maturity, while the human baby changes with an amazing speed—in fact, a rate of growth comparable to that of the embryo—for up to a year, then slows to a rate similar to that of "normal" postembryonic development. If we look at ourselves as mammals among other mammals, therefore, we see that we should be born twelve months later than we are; it is then that we emerge like other advanced mammals into the world of our peculiar kind, that we take on full human nature.

What is this "full human nature"? Portmann has mentioned its three chief characteristics: upright posture, speech, and reasoned action. All these have to be *learned* by the infant in its first months through contact with human adults or, in particular, with one adult, its mother or foster mother. Of course, parental care and social life in general are essential, in many intricate and marvelous ways, to a host of other species as well. But nowhere else in living nature, so far as we know, does just this pattern of premature sociality occur. Looking at it from the perspective of evolution and of comparative development, Portmann calls this unique period of postembryonic, yet still embryonic, growth the period of *social gestation.* In other words, to take its place alongside a newborn orang or chimp at an analogous level of development in relation to the adult form, the human being demands not only nine months in the physical uterus of the mother but a further twelve months in the social uterus of maternal care.

We must be careful, Portmann warns us, not to take this recently developed pattern in evolution as a mere addendum to an otherwise "ordinary" mammalian ontogenesis: our unique pattern of development is not an afterthought tacked on to a standard embryogenesis, as classical recapitulation theory would have supposed. The human attitudes and endowments that we must acquire in infancy are prepared for early in embryonic growth. Thus the first preparation for the upright posture, in the development of the pelvis, occurs in the second month of the fetus's growth. The preparation for the acquisition of speech, moreover, includes glottal structures very strikingly different from those of any other species. And the huge size of our infants relatively to the young of apes (who are born more "mature" but much smaller) is probably related, Portmann conjectures, to the immense development of the brain necessary for the achievement of human rationality, a development that begins very early in ontogenesis.

In short, the whole biological development of a typical mammal has been rewritten in our case in a new key. The whole structure of the embryo, the whole rhythm of growth, appears to be directed to the emergence of a culture-dwelling animal—an animal not bound within a predetermined ecological niche like the tern or the stag or the dragonfly or even the chimpanzee but, in its very tissues and organs and aptitudes, born to be *open to its world,* or, better, open to a world within a world, to be able to accept responsibility, to make the traditions of a historical past its own, and to remake them into an unforeseeable future.

Not only the first year of life, moreover, but the further pattern of human development testifies to the fact that, even in their physiological foundations, our lives are ordered so as to facilitate social learning—the perpetuation and modification from one generation to another of traditional lore. The postponement of sexual maturity to a date relatively later than is characteristic of other species permits a long period of apprenticeship and the gradual assumption of responsible personhood. And at the other end of the time span, the prolonged and gentle slope of human senescence provides opportunities for rising generations to profit from the richer experience of the old and wise. Thus at every level and at every stage of our existence we live out the uniquely flexible, uniquely creative pattern of configured time (as Portmann puts it) that is a human being.

From this vantage point we can see how various are the modes of life, including social life, of many, many kinds of living beings. We can also see how remarkable our own scheme of life is—unique among these uniquenesses in its "natural artificiality" (see next section, in the dimension of cultural indwelling that is not only added to, but enmeshed in and expressed through, every level of the many levels of order that converge to produce a human life.

With its talk of what our ontogeny and development "are directed to," what we are "born to," and the like, Portmann's language suggests an evolutionary perspective akin to that of Bolk and alien to our more Darwinian tradition. Yet, as Gould admits in his reference to this work, the description of the condition we have arrived at—never mind how we have arrived at it—is accurate and compelling. It is the social life built on this foundation that we want to put, in our terms, into its evolutionary context. For the moment, however, invoking yet another European source—the discipline of so-called philosophical anthropology, and, in particular, the version of it put forward by the sociologist and philosopher Helmuth Plessner—may help to consolidate the view of human sociality we are trying to develop.

Schema for a Philosophical Anthropology

Although concern about this seeming problem may be fading, biologists, and evolutionists in particular, have often tried to distinguish the subject matter of their discipline from that of the "exact"
sciences. Some put the distinction in terms of life as a history,
against the lawlikeness of inanimate phenomena (Gould 1986); some,
in contrast, see biology as richer in regularities than physics and
chemistry (Simpson 1961); some look to the nature of "teleonomic"
programs to make the difference (Mayr 1974, 1988). On a purely
descriptive level, the philosopher and sociologist Helmuth Plessner
characterized the difference between the nonliving and the living in
terms of the concept of *positionality*. Living things, he argued, set a
boundary between themselves and the outside world. They have a
boundary that at the same time *belongs* to them. He distinguished
between the open positionality of plants and the closed positionality
of animals. An animal, he wrote, has a center out of which it both *is*
and *has* a body. In the original German, the distinction between
Körper (just plain body) and *Leib* (organic body, or perhaps lived
body) helps to make the point. So far, so good. We are animals, too,
of course. What could be different about us? Through the curious
circumstances of our evolution, Plessner explains, we have come to
exhibit a variant on the closed positionality that all animals share.
Characteristic of human beings, he holds, is *eccentric* positionality:
the fact that, both being and having a body, we can view this duality
and deal with it. It is this capacity of standing aside, so to speak,
from our own bodily being that marks us off from other animals.
They live their bodies, as it were, naively, out of a center. We
discover our own centeredness and can take a position with respect
to it. (Plessner 1928; Grene 1974.)

On the basis of this description, further, Plessner formulates
three anthropological principles, two of which have an important
bearing on the interpretation of human social life.[4] First, there is
the principle of *natural artificiality:* uprooted from animal naiveté,
our natures demand the constructs of culture to support our daily
lives: ritual and customs from birth to death, as well as the more
tangible artifacts of clothing and shelter. Food, too, takes on artificiality of a sort through selection of ingredients, planting, cooking,
and serving in accustomed ways. All this suggests Rousseau's reasoning about the foundation of society: we need one another in
order to shape a place for ourselves within a nature that would

defeat each of us alone (Rousseau [1762] 1973). It recalls also, perhaps, Hume's treatment of the "artificial virtues" (Hume [1739] 1968). We have supplemented or even transformed the "limited generosity" on which our fellow feeling with others (and hence our moral judgments) is based with norms such as justice that can be constituted and maintained only within the boundaries of a human, contrived society. The "natural law" that underlies the constitution of human societies (if differently in each) is unnatural in that it depends fundamentally on the existence of conventions that constrain our behavior and even our desires. The artifacts of culture, far from being mere addenda to our natural lives, fulfill a deep need of our very natures.

Second, and closely related to Plessner's first principle, is *mediated immediacy*. We could say, perhaps, that the principle of natural artificiality brings into view a motif in the ontology of being human, while the second principle has a more epistemological significance: it pinpoints the way in which we apprehend reality. We sense the world directly, as all animals do, but even our immediate perceptions are mediated by the symboling activities of a human culture. Again, from the very early time when human infants are mature enough to recognize the expressions and the enunciations of their caretakers, they learn to *perceive* through the mediation of linguistic and other cultural devices. From the beginning, a baby's seeing and hearing are saturated with human meanings, and the meanings of natural objects, what they afford the perceiver, are immeasurably enriched by the human texture in which they come clothed. Perception itself is enculturated. Naming of objects and of pictures of objects may be almost simultaneous, and children may learn the names, for example, of zoo or farm animals, before they see those animals themselves. Thus our direct perception of the things and events around us is mediated by human customs and devices. Indeed in many, perhaps most, cases our environment itself is largely contrived, so that our perception is doubly mediated. On the one hand, our perceptual activity is aided by language and pictures, and on the other, the very objects of our perception are often the products of human interference with nature. Of course the latter point may be made about domestic animals, too: an urban dog hasn't much to see besides houses and streets, and what it smells, one supposes, does not much resemble what was familiar to its feral forebears. To pursue that point, however, would take us into the whole question of the humanization or denaturing inherent in domestication. All we are seeking to do here is to present Plessner's thesis about the mediation of our immediate awareness of our surroundings by the artifactualities of culture, both material and behavioral. Together with our need for artifacts, this principle marks the difference of human sociality from that of other primates, or other mammals.

It should be emphasized, however, that in stressing the principles of natural artificiality and mediated immediacy, in addition to the basic concept of eccentric positionality—the human ability to distance ourselves from our bodily being and bodily situation—we are not trying, in the fashion, say, of Boyd and Richerson (1985), to *add* a variable of cultural determinism to that of biological determinism. Adding a new variety of linear causality to that of ultra-Darwinian explanation gets us nowhere; this is sociobiology with a new wrinkle, no more. What we have to notice is that our dependence on culture, or, just as fundamentally, the potentialities opened up by our culture-dependent life-style, do not simply add a new causal line to a straightforward, one-level narrative. Our dependence on culture, our containment in culture, at the same time that it sets new boundary conditions for our behavior, gives us a new home (Merleau-Ponty 1942). As human beings we are from birth *in* the culture we are born, or adopted, into. That we live *in* language is but the most evident expression of this deeper and broader indwelling. Of course every animal exists only in an appropriate ecosystem; we too, as we are daily learning to recognize, depend on a natural habitat for our very being. And social animals in particular live not only in geographical territories but also in the relations prescribed by their social systems: in pair groups, troops, or what you will (what natural selection willed?), in special places (imitated also in human social systems, of course) in social hierarchies. Leo Buss has put forward an account of the evolution of individuality that stresses the *enclosure* of organelles or other subsystems within an individual (Buss 1987). Social life in general, we would suggest (despite Buss' emphatic disagreement; Buss, pers. comm., 1990) produces new enclosures, though short of producing new individuals in his sense. In vertebrate social systems, at least, the already evolved "individual" retains its status as such. Yet in social systems it is nevertheless enclosed in a historical entity that helps to shape its individual life. In the human case, further, that enclosing entity, rather than being species-specific, is itself engendered and maintained by the enactment of the conventions of a particular culture. Yet, in each case it is a real entity within which, in his(her) postnatal (and to some extent even prenatal) development, the particular infant-youth-adult has his(her) being and to which (in symbiosis with, often also in competition with and opposition to, his(her) cultural partners) he or she also contributes through individual constitutive, supportive, or even rebellious acts. Somewhat—though differently—as the skin of any animal encloses it and sets it as one individual over against the outer world, so social systems provide, in a manner of speaking, a second skin within which social animals exist, and culture, from this perspective, appears to provide a third enclosure, variable within the species yet historically and, in its expression, biologically and physically indubitably real in the way in which, as we have recurrently argued, historical entities are real.

We have developed, in earlier chapters, a hierarchical model of biological systems in which two hierarchies, the genealogical and the economic, are distinguished. Our view of animal, and in particular of human, social systems also reflects such a hierarchical perspective. But what kind of hierarchies are we invoking here? We are not saying that cultures, let alone states, are superorganisms. Yet there is a relation here between lower and higher levels analogous to that of biological systems. Human beings need a society, with its language, customs, and rituals, in order to develop into responsible agents, able to envisage and, other things being equal, to actualize relatively long-term goals. A human society, on the other hand, exists only through the existence of its members. Human societies may also involve complex relations of levels of organization within and alongside numerous subsocieties, families, clubs, churches, schools, museums, and so on. Such organizations in turn are variously related to the economic or the reproductive (or the sexual) aspects of biological reality (we will return to that theme in a moment). But it is a loose relation, in the sense that many such entities are only inadequately "explained" through their rooting in reproductive or economic activities. While kinship systems clearly function to control the vagaries of sexuality for the sake of effective reproduction, and while major-league baseball has become a plainly economic domain, it is hard to see how biological more-making or biological energy exchange necessitate or control the existence of, say, a local choral society or a given school of painting or style of architecture or direction in scientific research. Perhaps we are seeing here a reflection of Simon's insight: any complex system can come into being and maintain itself only if it is hierarchically organized

(Simon 1962). And of course we are dealing here with historical entities, of a kind that need biological individuals to originate and compose them; but given the efflorescence of culture, their tie to their biological foundations may be fairly loose.

The Eternal Triangle: Sex, Reproduction, and Economics in Human Social Organization

Given our schema for a view of human sociality, can we make a little more explicit the ways in which the themes of reproductive and economic adaptations have developed in the human case? Because vertebrate organisms all lead economic lives and engage in sex leading, in a majority of lives, to successful reproduction, it follows that vertebrate social systems display an inherently different form of complexity than do insect social systems. Complexity in eusocial insects is measured by the number of discrete morphological castes and enumerable chronological economic roles developed. With humans and other vertebrates, because there is a far wider range of activity for each organism, complexity lies in the myriad implicit among-organism interactions: there are economic consequences of reproductive behavior, as well as reproductive consequences of economic behavior. In humans, there is yet another complication. Here, sex has been decoupled from reproduction, if not further than in other primate species, at least in such a way that its liberation from the claims of reproduction seems to have given it a unique status in our lives, distinct from, but interacting in complex ways with, reproductive and economic activities. Thus in our case the social fabric is many times more complex still: there are economic and reproductive consequences of sexual interactions, as well as sexual consequences of economic and reproductive behavior.

Indeed, some recent authors (especially Wilson 1980 and Fisher 1982; see also Harris; 1989) have argued persuasively that acquisition of year-round sexuality (involving primarily loss of distinct estrus in hominid females) forms the very basis of the pair-bonding central to human social organization. Granted, this claim must be qualified in the light of the now acknowledged fact that sexuality has already been generalized to some extent in all the great apes (see chapter 6). But, as we have just suggested, the split in our case has perhaps gone further, or at any rate has cooperated with other features of our life-style in such a way as to call for a novel form of social organization.[5]

For example, Peter Wilson in his *Man: The Promising Primate* has attempted to ground the oddities of human sociality in a Darwinian theory of evolution (while refusing to explain them away in the manner of classic sociobiology) (Wilson 1980). Like others, he sees the pervasiveness of sexuality, and in particular the conflict between the primary bond of mother-child and the pair-bonding of sexual partners, as the phenomenon that demanded a new and artifactual solution. What our hominid ancestors first devised, he argues, was the invention of the father. Males mate with females but have no visible tie to offspring, who are bonded, natally and nutritively, to their mothers. With the invention of the father, then of the child as father's and mother's, institutions of human (rather than simple biological) kinship were established. It was within the conventional, yet controlling, hierarchies so initiated and maintained that we learned to make promises: to control the future, within limits, and to teach new generations to do the same—new generations that can also modify what their parents have taught them and teach new ways to new young. It is in these ways that human beings are "promising": while subject, like all living things, to the controls of sexual and natural selection, and, of course, to the vagaries of biological, geo-logical, cosmic contingencies, they have found ways to consolidate and perpetuate pair- and primary bonds through rituals of action and speech and through the practice of teaching. (Many animals learn; to what extent and how "purposively" do they teach? Culture, Wilson suggests, is taught behavior. That is perhaps too easy a slogan, but worth reflecting on.)

A feature of this new form of sociality that could be elicited from Plessner's anthropological principles—and also has its sources in a well-established literary-philosophical tradition—is the gappiness, or negativity, inherent in it (that is what the passage from Pico quoted above also reflects).[6] As Peter Wilson puts it:

> It must hardly occur to a giraffe during the normal course of its life that there is any difference between reaching up to the higher branches and getting leaves to eat. But for humans there is always a discrep-ancy between setting off on a hunt and coming back with meat; between planting seed and harvesting grain; between looking for berries and roots and finding them. (Wilson 1980:152)

Although this passage addresses an economic rather than a re-productive problem, and although Wilson deals, under the heading Dietary Considerations, with some economic matters, his chief argument, like so much Darwinian theorizing, is primarily related to reproduction—it is addressed almost wholly to the need for social control of a liberated and indiscriminate sexuality. Other writers, notably Fisher, have seen the "sex contract" as entailing economic benefits from the beginning (Fisher 1982). Her argument is not only that continual sexual activity is the precondition for stable cooperative pair-bonding, one that is essential for the successful rearing of offspring (given the protracted period of time required for human maturation), but that such pair-bonding always entails economic cooperation as well. Thus the social control of sexuality not only governs the production and maintenance of offspring but also provides cooperation in nutrition (game balanced by roots and seeds that females, even those carrying infants, can gather).

This basic conception seems an important one. As we have been arguing throughout this book, the economic as well as the reproductive aspect of biological interactions needs to be recognized in evolutionary discourse. We have been accustomed, ever since Darwin, to viewing the reproductive consequences of relative economic success in purely evolutionary, between-generation terms. It is, after all, the dynamic bias in reproductive success that Darwin called "natural selection." That modern evolutionary theory (unlike Darwin himself) has tended to downplay the economic (especially environmental) side of this equation in no way detracts from its fundamental importance. Not, of course, that we would deny for a moment that there are indeed reproductive consequences of economic behavior in humans as in other species; after all, the impact of economics on reproduction *is* natural selection. And natural selection has shaped human morphology and physiology, as well as, up to a point, our behavior, as assuredly as it has that of the organisms of any other species. What we are stressing, however, is the neglect of the economic aspect, which does after all take most of the time and attention of most animals. And in social vertebrates, including humans, the intertwining of the economic and reproductive activities of organisms that retain their individuality both as survivors and reproducers characterizes their diverse but species-characteristic social systems. In our case, this interaction is complexified by the separation of sexuality from reproduction; before we return to that theme, however, let us look at some of the problems that have appeared to numerous thinkers to follow from the reproduction-economics interaction in human societies.

Even before Darwin, Malthus worried about the apparent effects of relative economic success on reproductive rates in human beings; and if Malthus' argument in fact triggered Darwin's elaboration of his evolutionary insight, as Darwin claimed it had, this puzzling relation has lain at the foundation of modern evolutionary thought from its beginning (Malthus 1798; Darwin 1859). But it was apparently Francis Galton who first wrestled, in a eugenicist context, with the paradox that members of poorer, economically disadvantaged classes of large modern societies tend to outreproduce (per capita) members of the middle and upper classes (Kevles 1985). If, as the theory of natural selection tells us, it is economic success that provides the bias in reproductive success that leads to adaptive evolutionary change (usually construed as "improvement"), then humans seem to be in for a bleak evolutionary future indeed: especially if it is assumed (as it so often is) that membership in a particular social stratum is a direct reflection of ability, and, further, that such ability is heritable, then selection in humans seems (to eugenicists, that is) to be leading downhill, in the wrong direction. Human evolution seems to be in for a long spell of regression.[7]

More recently, the mayor of Singapore adopted the same point of view. His concerns were short-term: he was worried about the immediate socioeconomic future of Singapore, not the global evolutionary future of *Homo sapiens*. Yet the mayor stated the problem in precisely the same terms as had eugenicists before him: the poorer socioeconomic classes were outreproducing the middle and upper classes at a rate that seemed to him alarming. Nor was his concern limited simply to the threat of general overpopulation—the mayor was expressly worried that the number of poor people would increase at the expense of the middle and upper classes, with consequent general negative effects on the economic health of Singapore in general.

Economic success does not, in fact, have simple, deterministic, and automatic effects on reproductive success in humans (where, arguably, it *does* have such effects in other organisms, especially in asocial orgnisms); but this reflects the nature of the organization of social systems more than it bespeaks any necessary "decoupling" (in humans) between economic and reproductive success—or that humans have somehow entirely eluded the grasp of natural selection. And what the generally observed propensity of members of lower socioeconomic classes to reproduce at rates higher than middle- and upper-class members has generally been understood to mean (at least since Marx) hinges on the reciprocal relation that exists between economics and reproduction in human social organization.

Leaving a simple natural selection model, or simply a biologically derived model of factors influencing rates of population growth, and looking instead at the actual pattern of economic and reproductive interaction in human societies, we notice that children, under certain socioeconomic circumstances more than others, are viewed literally as a form of wealth. Reproduction, in this context, is the production of labor; poor families in society after society (that is, in post-industrial revolution Western societies—Singapore qualifying through its colonial past) typically produce relatively larger numbers of children per family both to ensure the survival of some to adulthood (mortality rates are typically higher in these families) and to secure economic labor in support of the livelihood of all family members. Even infanticide as a population control mechanism fits this pattern: infanticide occurs when children's presence poses a greater immediate economic threat than a potential economic (or reproductive) benefit in the long run. And, as Harris (1989) has stressed, infanticide typically involves a disproportionate number of females—presumably reflecting, as well, a difference in economic potential.

Meanwhile, the relatively greater economic success of members of the middle and upper classes seems to have the reverse effect; on average, they produce fewer children than do members of the lower socioeconomic classes. Although it has been maintained that the long period of childhood favors maximizing attention to smaller numbers of children—that is, it is *adaptive* to raise relatively few children well than more in a disadvantaged state—it could as easily be argued that children with relatively advantageous backgrounds have a greater chance for their own economic success in the future: the economic returns may well be greater with fewer children with superior advantages (especially education). But the effects of economic success in (apparently) depressing reproductive rate has had the result, in our own society, of producing DINKS ("double income, no kids," recently so delineated by Madison Avenue)—heterosexual couples, each partner of which has a career, and that have none of the financial and other obligations of child rearing.

We make no claim to have understood fully the underlying causality of such patterns. But it is clear enough that economic life impinges on reproductive life, and vice versa, in human societies—

and, moreover, does so in ways that are counterintuitive to a pure (ultra-)Darwinian. It is possible to look at average family life per socioeconomic stratum and to make the claim that number of children maximizes overall probability of reproductive success (fitness) of the parental generation. Yet Galton's cry does point to a paradox that can be fully resolved only in the realization that there are direct economic ramifications of reproductive behavior—implications that differ depending upon the economic circumstances of the parents. Even in eusocial insects, of course, there is a direct effect of reproduction on economics—most of the reproductive activity goes into the production of members of the working castes.

As we have already noted, Fisher (1982) has developed an image of (human) nuclear families (particularly, male-female stable pair-bonds) as hybrid economic and reproductive cooperatives. The bridge, in her model, is sexual behavior per se. Thus in human social organization, sex, as an item of behavior distinct from reproduction, provides many of the links between economics and reproduction. According to Fisher, the availability of year-round sex is the very precondition that keeps males around for life. Not that this is a necessary condition for long-term mating in every species. That continual, year-round sex is not a prerequisite for stable pair-bonding in general is shown, for example, by the otherwise asocial kingfishers as well as other birds that mate for life yet retain a distinctly seasonal sex-for-reproduction-only mode. In our case, it would seem from Fisher's argument, the cultural systems made possible, and necessary, by our neotenic development and consequent long juvenile dependence are sustained and stabilized by the use of generalized sexuality to weld reproductive and economic needs. Her "sex contract" is basically a "food for sex" arrangement: females gain economic advantages (improving their chances of obtaining hunted food) in return for their sexual availability. Of course, the arrangement is reproductive, as when offspring obtain not only maternal care but also a share in that same meat (and later in the scenario, other benefits of paternal as well as maternal care), and economic, insofar as the male supplements his diet with female-gathered nuts and berries.

Once the basic pattern is established, the list of potential interactions between the three variables—economic, sexual, and reproductive behaviors—turns out to be virtually limitless. Sex, of course, is necessary for reproduction. Yet reproductive activity affects sexual behavior, as most people with infants can attest. That sex and economics are linked is equally notorious—and labyrinthine. Sex is packaged and sold; and much economic behavior, especially within-sex competition in the workplace, is seemingly reflected directly in sexual activity in the workplace. Although such characteristics of our particular society are not generalizable as such, they presumably provide particular instances of relations between sexuality and economics that are expressed in other forms in other societies.

Similarly, we may look in our own society for examples of the relations between economics and reproduction, in addition to the Malthusian paradox. Complex human social systems consist of intricate series of hierarchically nested entities (organizations, political systems, etc.). Restricting attention for the moment to a special case in our current social system, to families depicted on Wheaties boxes of the 1950s and 60s (father, mother, two children, often one of each sex): such units live in cross-genealogical contexts; rarely does one family group live only in proximity to extended kin. Neighborhoods are distinctly cross-genealogical. There are no reproductive organizations to which members typically and habitually belong (beyond Lamaze groups, and medical care around the time of actual birthing: planned Parenthood, it can easily be maintained, is concerned directly with the economic effects of overpopulation). The many entities to which family members find themselves belonging (a class in a school, a patrol in a troop, a department in a firm, a committee or board of an organization) are virtually all economic, at least in the extended sense that they are constituted purely through energetic interactions rather than reproduction. Even if, as Fisher claims, a sex contract lies at the basis of stable pair-bonds, and even if such bonds lie at the heart of human social organization (as can be seen particularly in less complex societies than our own), still most human life and most elements of human social structure and organization are devoted to (and arise from) the purely nonreproductive, economic activities of humans.

Human social life is no more purely about reproductive success than it is purely about economic success—any more than it is in any other species—although if one were forced to pick the dominant element, especially in humans, the economic seems to be the most important. The invention of the father, if Peter Wilson is correct in his diagnosis, while permitting the foundation of human kinship systems, at the same time permitted the complex self-imposed obligations inherent in all the daily activities of members of a human society, which are largely economic, or quasi-economic, in nature.

Thus economics now seems to dominate the sexual and reproductive interests that the human form of life originally served.

If the nuclear family, which has clear ties to reproduction, is deeply enmeshed in economic activities, sex, too, of course, promotes social cohesiveness of some sorts of human aggregates. Homosexuals (as well as DINKS) can and do form stable pair-bonds. Homosexuality would seem to pose a distinct difficulty for ultra-Darwinians; if one assumes that the ultimate purpose for any sort of behavior, including economic, is reproduction (based on the correct perception that, all things being equal, differential economic success causes differential reproductive success), one must rely on kin selection (support, by homosexuals, of reproducing kin and their offspring) to explain how homosexuals maximize their fitness. In stable pair-bond situations involving homosexuals, the benefits are patently quality of somatic life through mutual support—it is an emotional relationship that has direct effect on the economic lives of the partners. It could, perhaps, be maintained that such stability and economic welfare only increase the chances of kin selective mechanisms' being enacted: vicarious reproduction. Indeed, that is in effect the sociobiological explanation of homosexuality, which it has assimilated fairly smoothly—as distinct from the liberation of women, which it cannot, or will not, deal with at all (Wilson 1978). Yet it is far more straightforward to maintain that human social organization, most of the time, is about economics (in the broadest sense), not about reproduction or even sexuality, and that even structures developed as economic/reproductive cooperatives, in which sex plays an important interactive, integrative role, can "secondarily" become purely economic, or economic-sexual (as in the DINK case), or be mimicked by same-sex pair-bonds, forming social entities of a strictly and frankly sexual and economic but not reproductive nature.

In short, economic behavior, though linked in complex ways to reproduction and to sexuality, on the whole dominates human existence, as it does that of almost every other organism (except, perhaps, male mosquitoes and other creatures [usually male] who live, like the male mosquito in its terminal imago stage, seemingly only to reproduce). Most human interactions—and especially the sorts of interactions that integrate social systems—are economic, or quasi-economic, in nature.

The fact that economic institutions are spatiotemporally historical entities throws the matter into still another light. General Motors and the American Museum of Natural History, to take two examples, exist, in a real sense, in their own right, and perhaps even for their own sake. People work in such institutions to earn a living; but such institutions are managed with a clear eye towards perpetuity. They depend, implicitly, on "promises" that transcend the particular needs of individual employees or staff. In other words, effort is expended to improve such institutions. Motives for making such efforts may well be continued employment or enhanced enrichment and reputation of the owners, trustees, and employees entrusted with the care of the institution, but the stated goal is the enhancement and continued existence of the institution itself. And, of course, such institutions live on as rather timeless, even stolid, entities, despite the lower-level turmoil of employees coming and leaving in a constant stream. Rather as the somatic cells of an organism's body are constantly being formed, living, then dying and being replaced by new ones while the body itself remains the same (or at least is on a different timetable of change)—the quintessence of organismic economic activity—people staffing large economic institutions are constantly being replaced with generally little immediate effect on the character of the institution as a whole. Human individuals work hard to help perpetuate such entities, which are potentially more permanent than their individual lives—in doing so, transferring their own economic activity to more generalized, nonbiological historical entities.

Coda: Cultures as Symbol Systems

The previous section glanced at aspects of the interactions of sexual, reproductive, and economic behaviors conspicuous in human societies. In particular, we stressed the importance of the economic factor, which is virtually neglected in ultra-Darwinian accounts, as distinct, of course, from the theories and descriptions of sociologists and anthropologists outside the idiosyncratic sociobiological tradition.

Finally, our last example recalls, and allows us in conclusion to touch more explicitly on, a theme suggested earlier: the dependence of human sociality, and hence of human life as such, on the enactment and maintenance of networks of symbolic activities and the social entities they mediate. Human language, as we said earlier, is the most massive and evident expression of this need and power. But ritual may silently, in dance or gesture, perform the same symboling function. And the artifacts we institute by such activities often acquire a quasi-life of their own. The self-perpetuation of an institution like General Motors or the American Museum of Natural History is metaphorical; it does not *biologically* come into existence or maintain, let alone reproduce, itself. It exists, of course, temporally from one year to another; but spatially and physically it could move its headquarters to a different habitat or abandon its old buildings for new ones; it also recurrently "consists of" a host of new individuals held together by its fabric of rules and customs. In a way it forms a metaphorical ecosystem, through ordered interactions within which its members form the social equivalent of avatars. Here—always within nature, within the boundary conditions set by physical and biological constraints—the genealogical, the sexual (in its liberation from reproduction) and the economic aspects of human lives are united in the complex networks of organizations and traditions through participation in which we become the persons we have the capacity, and opportunity, to become.

Thus human institutions, and traditions within or alongside or in opposition to such institutions, produce and are produced by the *social spaces* generated and supported by the activities of their participants. Again, such spaces are not wholly and sometimes not primarily literally "spatial." To take a less than institutional example, a charismatic individual need not be bigger, take up more room, than a less impressive person. At the same time, it is the bodily spacetime (Munn 1986) of an individual human being, a group, an institution, or a society that is constituted by its activities or by the activities of the entities that compose it. Physical events, such as a snowstorm, happen; human events *take place*. That dead metaphor has its significance. As Charles Taylor puts it, human beings have to *take a stand* somewhere in relation to a good or goods through which their lives become significant—become human (Taylor 1989). To be human is not just to be alive; it is to live within the bounds not only of nature but also of a given complex of symbolings and symbol systems—a given language and the enactments that support it.

A good example of the role of *social space* in the origin of a subculture is provided by Shapin and Schaffer's (1985) analysis of experimental methods in natural science as they developed in the circle around Robert Boyle in the late seventeenth century. Not only the actual places—Gresham College (home of the Royal Society) or the London residence of the Duchess of Newcastle, where some of Boyle's experiments were enacted—mattered in this story, but also the place of experiment and experimenters in British society, in its relation to political and ecclesiastical forces. Boyle's praise and implementation (through others' hands, be it noted) of experimental investigation of natural phenomena, particularly the invention of the air pump and the doctrines its use supported, were closely related to his conception of the correct social standing of subjects, church, and monarchy in a well-ordered state. And the opposition of Boyle and his group to Thomas Hobbes's diametrically opposite view of science and state was deeply involved in their rejection of his antiexperimental, more axiomatizing view of "natural philosophy." Shapin and Schaffer do, perhaps, go a little too far in politicizing the origins of experimental science. Within a given society, individuals and groups of individuals may develop—and in the case of natural science in the past three centuries in the West, have in fact developed—traditions and institutions whose social space depends primarily and directly (whatever the motives of many of their participants) on the passion for knowledge, which outruns (or falls short of?) the more directly biological aims dictated by economic need or reproductive drive.[8] Yet the sciences exist only insofar as they have developed their own characteristic social spaces. Not that such developments are usually as self-conscious or explicit as they were in Boyle's case; most social spaces are constructed on the level of practice, not theory. But the example is nevertheless instructive.

Let us return now to our previous examples, which will enable us to elaborate a little on our present theme. General Motors, of course, exists for purely economic reasons and is perpetuated by need—and greed (as well as by Hobbes' and Thucydides' other motives, fear and vanity; but it is the purely economic we were stressing earlier, and it is certainly canonical here). A service institution like the American Museum, however, is to a degree removed—as are the scientific institutions that had their beginning in such groups as the early Royal Society—from the immediate biological basis that supposedly supports the profit motive. In other words, its staff and employees may be concerned, at least in part, with the ends of education and enlightenment it is meant to serve. And the energy expanded by the massive collections of schoolchildren who visit it, although it relates indirectly to the "livings" of their teachers, seems to have little direct connection with the children's economic lives. One may, of course, argue that they will grow up to acquire better jobs and more worldly goods if they have an early acquaintance with dinosaurs. Whether in sociobiological, Marxist, or other reductionist terms, all cultural realities *can* be reduced to the terms of lower-level entities and processes. But such reduction carries with it blindness to the many dimensions of human life made possible by the symbolic activities that constitute and maintain human social systems.

Given both our immaturity at birth and the pervasiveness of a sexuality divided from reproduction, in order to survive and reproduce we have had to devise systems by which we symbolically structure our largely unordered potentialities. Once such systems exist, however—let us insist once more—they not only confine our aims and activities, but they also give them space to develop. Our pair-bonding, our need to procreate, and our expenditure of energy in response to individual needs, linked together as they are in complex and various ways in a given culture, can be transformed in such a way that a given embodied human being is shaped in his (her) life history by participation in (bonding to) institutions, traditions, transcendent goals, and goods that far outrun the reproductive and economic adaptations that more remotely underlie and limit such commitments.

The Freudian term *sublimation* comes to mind here, as does Plato's ladder of love in the *Symposium*. From binding to and hoping to procreate with one body and then one soul, a human person can come to love—to seek union with the good—through laws, institutions, or other entities wider than individual need, whether sexual, reproductive, or economic. It is this transcendence of individuality *in* the bodily human individual that makes possible the major achievements of human culture in the arts, science, law, and morality.[9]

There is a literature of social anthropology, in particular symbolic anthropology, that explores such phenomena in a number of diverse cases (see Ortner 1984 for a recent review). It would be inappropriate to embark here on an exploration of that or other analogous material. Let us just illustrate, in conclusion, one attempt to describe one facet of human social practice—myth making—for myths (akin to what biologists call models) typically support the social entities through participation in which, and within which, members of a given culture develop. Pierre Bourdieu, an influential writer in this tradition, whose fieldwork concerned the Kabyle people of North Africa, says of myth and myth making:

Every successfully socialized agent thus possesses, in . . . [his/her] incorporated [that is, bodily] state, the instruments of an ordering of the world, a system of classifying schemes, which organizes all practices, and of which the linguistic schemes . . . are only one aspect. To grasp through the constituted reality of myth the constitutive moment of the mythopoeic act is not, as idealism supposes, to seek in the conscious mind the universal structures of a "mythopoeic subjectivity" and the unity of a spiritual principle governing all empirically realized configurations regardless of social conditions. It is, on the contrary, to reconstruct the cognitive and evaluative structures which organize the vision of the world in accordance with the objective structures of a determinate state of the social world: this principle is nothing other than the *socially informed body.*

And that means the individual, lived body of an individual human being, "informed," that is, given order, structure, and direction, through the society into which it has been cast "with its tastes and distastes, its compulsions and repulsions, with, in a word, all its *senses*, that is to say, not only the traditional five senses—which never escape the structuring action of social determinisms—" (thus, as we noted earlier, the very bodily perceptions of a human infant are in large part already shaped by the mores, what Bourdieu calls the *habitus*, of its society) "but also," he continues,

> the sense of necessity and the sense of duty, the sense of direction and the sense of reality, the sense of balance and the sense of beauty, common sense and the sense of the sacred, tactical sense and the sense of responsibility, business sense and the sense of propriety, the sense of humour and the sense of absurdity, moral sense and the sense of practicality, and so on. (Bourdieu 1977:123–124)

This is, if you will, a lot to hang on a wordplay with the two meanings of sense (or three, in French). But the passage does suggest, cryptically, the many dimensions in which the narrative that constitutes a given bodily human life is woven into the tissue of complex cultural patterns—patterns that in turn acquire, in part metaphorically, an existence and a history of their own, sustained by the activities of those very beings who, reciprocally, depend upon them for the significances that make their lives characteristically human. It is to be stressed, finally, that while myths, like other major artifacts of culture, are, in Geertz's term, "factions," they are nevertheless not fictions (Geertz 1988). They are the paths we make, and take, to reality—always within the contingent and even paradoxical situation, itself a containing, supporting, and limiting reality, that our phylogeny has bequeathed to us.

In conclusion, then, we have tried to sketch a view of human society that stresses both its likeness to and difference from other vertebrate societies. Our sociality, like that of other vertebrates, is characterized by the interaction of reproductive and economic factors, with the nearly independent development of sexuality forming a complicating third ingredient. Unique in our case, in particular, is the dependence of human beings on the artifacts of a given culture, which, though "invented," contains its members as surely as other animals' species-specific social organizations enfold and organize their members' lives. Thus we have been adding here another example of the theme we have been developing throughout. It has three parts. First, we have looked at the situation in evolutionary biology, both currently and in its history, and have found a need for, and in part already a realization of, a broader and more complex theory than traditional Darwinism (especially ultra-Darwinism) provided. Second, we have argued, the supplement evolutionary biology most needs is to take account of two major hierarchies cast up by the history of life on this planet: a genealogical hierarchy, itself fundamentally historical, and an ecological or economic hierarchy, which taken on its own is relatively synchronic: a form of organization rather than a history—although, of course, natural selection both depends on economic differences to occasion the differential reproduction with which many identify it, and, through differential reproduction, produces novel economic systems. Third, social systems result from the reunion of distinct genealogical and reproductive phenomena or behaviors (though differently so in colonial, insect, and vertebrate cases). Such systems, in turn, have to be counted among the historical entities whose reality an adequate biological ontology must recognize.

Notes

1. Contemporary Evolutionary Perspectives

1. Grant's *Evolution and Ecology of Darwin's Finches* (1986) forms a notable exception, since he does explicitly consider problems of speciation as well as of adaptational differences between or within extant species.

2. Ernst Mayr has criticized hyperadaptational programs (presumably including sociobiology) for their atomism, but suggesting in place of our "pseudoteleogy" a misplaced holism (Mayr 1983, 1988). The tension between atomism and holism, as we argue in chapter 2, is indeed characteristic of the Darwinian tradition, but it is difficult to see what is "holistic" about so clearly reductionistic a theory as sociobiology.

2. Darwinism and Ultra-Darwinism

1. We are grateful to Dr. Daniel Kolb for pointing out this passage to us.

3. Entities, Systems, and Processes in the Organic Realm

1. It is perhaps unwise, in a book about the biological context of social behavior, and in particular human social behavior, to use the word "economic" in this particular broader sense—that is, to mean processes of matter-energy transfer in biological systems, be those systems cells, organisms, populations, or ecosystems. Yet it is just this distinction between a

general class of essentially economic behaviors and those involving the maintenance, modification, and transfer of genetically based information—genealogical—processes that forms the basis of our analysis. In any case, the term *economic* has been used in this sense extensively in recent years (see Eldredge and Salthe 1984; Eldredge 1985, 1986).

A further possible source of confusion is Ghiselin's (e.g., 1974b) insistence that reproductive activities be viewed, in an evolutionary context, as an aspect of the "natural economy."

2. Thus reproduction is usually said to be "for the good of the species," a formulation looked on askance by good Darwinists, since natural selection, strictly an among-organism/within-population, generation-by-generation process, cannot possibly build adaptations that are neutral or harmful to organisms (reproduction takes energy) but serves instead some imagined benefit to a large-scale unit—such as an entire species. Thus sex—even more wasteful (arguably in terms of energy output, and certainly in terms of efficiency of transmission of an organism's genetic information)—is widely seen as a conundrum for traditional Darwinism (for further discussion, see chapter 4). But the entire process of reproduction, of whatever mode, poses the same general problem. It is not until we realize that protoplasm is mortal—in fact, and probably in principle—that we see that selection at its crudest is responsible for the very existence of reproduction. The ubiquity of genetic transfer and lability in early biotic systems (see Margulis and Sagan 1986) probably ensured that reproduction in some sense had to exist; were it possible to have organisms that do not (because they cannot) reproduce, they would have become extinct almost as soon as they arose. We need not argue that reproduction was invented to avoid extinction; but because it was a part of the chemical physiologies of early biotic systems, life has survived as long as it has—as an incidental side effect of reproduction.

3. Viruses need the cells of other organisms both to reproduce and to manufacture proteins; thus viruses, structurally more simple than bacteria, are not independent forms of life and must be secondarily parasitic.

4. Yet even within Protista some semblance of multicellularity appears, as with the photosynthetic colonial flagellate *Volvox*. See chapter 6 for further discussion.

5. Acoelous animals form yet another heterogeneous array of have-nots—all of which, in this instance, lack a true mesodermally lined coelom, or body cavity.

6. Again, we are speaking here of nonsocial organisms; examples from within the set of social organisms are numerous, such as the recently designated DINKS (double income, no kids), targeted for economic reasons, presumably by Madison Avenue. We return to DINKS in chapter 7.

7. Patterson's Specific Mate Recognition System (SMRS) involves far more than behavioral and anatomical aspects of mate choice—which itself may be based on criteria of overall robustness. Rather, the SMRS involves

(and is limited to) *all* aspects of the phenotype that pertain to species-specific *reproductive* phenotypic attributes, including chemical "recognition" of conspecific gametes in all sexual organisms, as well as the more obvious behaviors and morphologies that allow conspecifics to recognize one another. It is the SMRS that sorts out conspecifics, thus appropriate mates, from nonconspecifics.

4. Biotic Consequences of Organismic Reproduction

1. For example, species composed of organisms that are relatively broad-niched (eurytopic) as a side effect seem to display characteristically greater longevities when compared to more narrow-niched close relatives; see Eldredge 1989 for discussion and references to this phenomenon.

2. Note that Mayr in 1942 had two different conceptions of species on facing pages (152–153) of his *Systematics and the Origin of Species*. Species were reproductive communities that arose by fission from ancestral species—and were analogous to *Paramecium* individual organisms. But, through time, species are arbitrarily designated segments of a continuum of evolving modification within single lineages: Mayr supposed that, were the fossil record complete, a paleontologist would not be able to see discrete species, but would be forced instead to chop up a continuum of anatomical diversity into separate species seriatim. These sorts of entities are usually termed *chronospecies*. They are collections of similar organisms within lineages, rather than imagined as single, coherent communities reproductively isolated from ancestral and descendant species. For more on this aspect of Mayr's species ontology, see Eldredge 1985, especially chapter 3.

3. These issues pertain also to mechanisms of speciation, that is, the origin of descendant from ancestral reproductive communities. Perhaps the most notable example is Dobzhansky's (1937) formulation of the principle of "reinforcement," whereby reproductive isolation is seen in part as a product of selection working against hybridization—so that the two fledgling separate species can better focus their economic adaptations to their different adaptive peaks. See Eldredge 1989, chapter 4, for further discussion.

4. The standard model of allopatric speciation simply invokes accrual of genetic differentiation leading eventually to inability of full, unimpeded reproduction to occur. No distinction is made between economic and reproductive features; indeed, it is usually assumed that economic divergence leads to the accrual of incompatible behaviors, morphologies, and, ultimately, genetic incompatibilities that actually underlie speciation. While it is likely that such processes do occur, Paterson's notions, followed here, place far stronger emphasis on simple disruption of SMRS directly (i.e., not as a secondary outgrowth of general selection- or drift-mediated change in any non-specified set of organismic features) as the primary ingredient of speciation.

5. Organismic Economics and Biotic Organization

1. Oster and Wilson (1978) have coined the term *ergonomics* essentially to mean (biological) economics in the sense used here. It is tempting to adopt their term, especially in view of the latent confusion between biological and frankly *human* economics (i.e., economics in the conventional sense). We avoid temptation, however, because *ergonomics* has not gained wide usage in the decade since its coinage.

2. Wynne-Edwards (1962, 1965, 1986) is the best known of the comparatively few biologists who posit factors other than resource limitation (and other external environmental elements such as climate, predation, and disease; cf. Darwin 1859) as the controlling factor of population size. Wynne-Edwards instead argues for internal self-regulation of population size, maximizing the overall fitness of the group rather than the fitnesses of individual organisms (as when economic success spills over into reproductive success—the standard model of natural selection at the heart of the resource limitation model of population size). Wynne-Edwards's model is geared explicitly to social animals, though, of course, all organisms experience limits to population size.

3. Colonies, of course, are not individuals in the same way that single organisms are individuals. Indeed, individual organisms within colonies are not individuals in the same sense as are free-living organisms. See Buss 1987 and Eldredge 1985 for further discussion of these points; in any case, the distinction between kinds of organisms is not germane to the present discussion. We return to colonies as a special type of social system in chapter 6.

4. We refer here to nonsexual interaction, though economic and reproductive functions might simultaneously be served by a single item of behavior, as when a male tiger marks the boundaries of his hunting/mating territory.

5. Vrba (1988) and Eldredge (1989:86) have commented on the ontological confusion in Stenseth and Maynard Smith's analysis—specifically, their treatment of evolutionary patterns in populations (avatars) within ecosystems as if they were the same as patterns within species as whole entities. However, our point here is only that some biologists have claimed not only that a spectrum of degree of accommodation within ecosystems (or communities) exists, but also that the spectrum has implications for evolutionary processes and resultant patterns.

6. Conservationists, including conservation biologists, are very much interested in the histories—and futures—of ecosystems. The rhetorical question posed here is meant to broadly characterize the difference seen in the importance of history in evolutionary versus ecological biology over the last century or so.

7. Pilette (1989) has recently argued that use of the term *organism* to designate the presence in both hierarchical systems of entities of different

kind, albeit representing the same level, is to be avoided. He therefore adopted the term *genome* to represent the organismic level in the genealogical hierarchy and the term *phenome* to represent the organismic level in the ecological hierarchy. Though there is some appeal to this suggestion, the fact that genome and phenome are housed within the same unit (in other terms, Dawkins's replicators are a part of the vehicle), that replication/more-making and economic activity are largely segregated parts of a single entity, an organism, should be recognized in the formal levels of both systems. In other words, we see no particular difficulty in the fact that the organismic is the only level in which the same term, *organism,* appears in both hierarchies.

6. The Biology of Sociality

1. Wilson (1975) treats metazoan invertebrate colonies as forms of social systems. Plants do, of course, develop comparable systems through asexual (vegetative) reproduction. Though there is a division of economic and reproductive labor in such systems, there is little, if any, further economic division of labor comparable to that encountered in marine invertebrate colonies, as we discuss later in this chapter.

2. Examples of eukaryotic cells adhering to form a colony, with some integration but little or no differentiation, may represent vestiges of the actual phylogenetic origin of metaphytes (in the case of *Volvox,* i.e.)—or simply an independent move in the direction of true multicellularity. In either case, we take *Volvox* to represent a general level, an exemplar of a primitive stage of the development of true multicellularity.

3. Actually, of course, there are many levels of individuality here, as in any biological system. For example, cells do not lose their status as individual entities simply because they are aggregated to form larger systems (tissues, organs, entire multicellular organismic somas). We allude here only to the discussion, frequently posed as an either/or problem, of "what constitutes the individual in a colonial system formed from the association of what could be/might have been separate organisms?"

4. Yet it is Oster and Wilson's (1978) devotion of the bulk of their discourse to *ergonomics*—the economic activities, especially the caste systems and division of labor in eusocial insect societies—that may well constitute their signal contribution to understanding the biology of social systems generally. It is clear that economic behavior (ergonomics) is, in their view, subservient to reproductive activities—which constitute the sine qua non aspects of any social system. Nonetheless, in our arguments herein attempting to redress the balance (i.e., in perceived significance) of economic versus reproductive behavior, we do not wish to imply either that (1) reproductive considerations are *not* to be taken as important to social organization or that (2) sociobiologists never consider the economic side of the ledger. It is the relative emphasis of the two over which we differ with sociobiology.

5. A definite but distinctly muted variant theme in sociobiological literature that many authors have acknowledged is that social organization has immediate and direct benefits to component organisms—enhanced protection, securing of resources through cooperative hunting, and so forth. Although competitive maximization of genetic representation in the next and succeeding generations is the dominant sociobiological answer to the *why* of social organization, this is not to say that biological theorists addressing social organization have ignored direct economic advantages to components. However, we reiterate that immediately upon presentation of such a list of direct, economic benefits of social organization, the analysis almost invariably turns to the supposed deeper structure—the genetic advantage of such organization.

6. Naked mole rats (Sherman, Jarvis, and Alexander 1991) appear to constitute the only true exception to this generalization. Indeed, with the single reproducing female (queen); sudden growth and reproductive status of a replacement when the queen dies; one or at best a few reproducing males; and development of particular roles for worker rats (i.e., all the remaining rats of the colony), the social structure (including, especially, the division of labor) of naked mole rats is strikingly like that of eusocial hymenopterans—and unlike that of other vertebrates.

7. That adult male elephants live as lone bulls, quite apart from such female and young collectivities for much of the year, highlights, as well, the distinction between demes and avatars (the reproductive and economic, respectively, aspects of local populations) drawn in earlier chapters.

8. Evolutionists (e.g., G. C. Williams 1966) often require, for the demonstration of modes of selection to be operating at levels other than the among-organism level of natural selection, that the results of the putative selective processes be shown to be in opposition to one another. Williams (1966) had group selection in mind: to demonstrate group selection, that is, to show that a mode of selection other than natural selection is in operation, it would have to be shown that the results obtained were the opposite of what would have occurred under a purely natural selective regime. So, too, it might be argued, with natural and sexual selection: sexual selection (in Darwin's sense: the advantage that organisms have over others of the same sex and species, purely in relation to reproductive success) favors (in foxes) defensive protective care of young; natural selection, however, otherwise encourages far more cautious, secretive behavior when identical threats (such as encounters with humans) occur. In reproductive defensive behavior, the female fox takes a much greater risk with her own economic life than she otherwise would in nonreproductive situations.

9. In this context, evolutionists have clearly overstated the nature of evolution *as a process*. There are processes of matter-energy transfer and of reproduction. There are consequences to these activities, namely the formation of various sorts of stable entities, and biases in the distribution of

genetically based information arising from interactions of various sorts among such entities. It is the latter, strictly speaking, that is "evolution"— the fate of genetic information. Evolution is passive bookkeeping, a differential bias reflecting what worked best in previous generations. It is not *for* the spread of genes but is, instead, a reflection of the ingredients of past success—most of which is in the purely economic sphere. It seems obvious that evolutionary biology has confused the accumulation of history with the dynamics of actual process: the moment-by-moment economic and reproductive processes of organisms, their components, and the larger-scale entities that form as a result of such activities.

Hagen (1989:447), it is interesting to observe, makes nearly the opposite point. He claims that treatment of biological entities as structural/functional systems (he had ecosystems specifically in mind) should at least be consistent "with accepted evolutionary theory"—a reasonable enough notion, unless we take current evolutionary thinking as an unalterable benchmark against which all functional descriptions of biological systems are to be judged. The dialogue between proponents of functionalism and those of "evolutionism" is indeed an old one; its latest manifestation comes from a renewed interest by developmental biologists to shed (or at least loosen) the shackles of evolutionary analysis in the interpretation of ontogenetic data (cf. Goodwin 1988).

Clearly, all biological systems have structure and display functions; they all have histories as well. In a nutshell: ultra-Darwinian evolutionary theory emphasizes the generation of history but pays rather little attention to the ongoing, dynamic, structural and functional aspects of the systems the histories of which the theory is geared to explain. But emphasizing the structural and functional aspects of biological systems—their modes of organization and the types of work, if any, that they perform—should by no means exempt any of us from considering the manner of production of their evolutionary histories as well. The present work is an attempt to inject a bit of structuralism/functionalism into evolutionary biology generally, not an attempt to pit the one against the other. And we are addressing social systems in particular—an area where evolutionary considerations, based on a particular view of the nature of the evolutionary process, are said to govern all elements of social structure. It is here that a view of social systems arising as stable entities from the interaction between economic and reproductive activities of organisms could, in our view, serve as a useful complement to a purely historical, evolutionary point of view.

7. Human Sociality

1. At the species level, an SMRS, as a complex of characters, may perhaps be taken as *one* property that marks a species.

2. It is important to point out that this Kantian maxim need not be interpreted in the idealistic sense characteristic of modern neo-Kantians, especially in France. The constitutive work of human agency referred to

need not be attributed to some kind of mythical "self-consciousness." See the criticism of such neo-Kantian readings in Bourdieu 1972.

3. Much of the report of Portmann's work is taken with some revision from Grene 1974, which is, in turn, based on a careful study of the original texts; it seems unnecessary at this juncture to change that account in any substantive way.

4. The third principle, that of the "utopian standpoint," appears less fundamental and, indeed, rather strained. It has in any case no direct bearing on our argument here.

5. Even Savage-Rumbaugh, who wants pygmy chimps to do everything we do, admits that our social organization is distinct from theirs (Savage-Rumbaugh et al. 1979)!

6. Wilson's introduction gives a nice survey of the Enlightenment conceptions of the paradox of human sociality, running from Hobbes to Kant, and even Marx.

7. For an account of the human inhumanity consequent on acceptance of the eugenics program, see J. David Smith 1985.

8. Rita Levi-Montalcini describes the last hours of her teacher, Giuseppe Levi, as follows:

> Our farewell—in that small hospital room whose bareness reminded me of a Franciscan's cell—could not have taken place in surroundings or under conditions more apt to bring out his qualities as a man and a scientist, consumed neither by old age, nor suffering, nor by the knowledge of his approaching end. He accepted the interest in research as an instrument for the understanding of nature and not as an object of competition and an instrument of power. In an era when the latter conception of scientific activity prevails, that last conversation of ours, the vivid interest he took in what he knew must be his last "briefing" on the state of my research, revealed to me the secret of the great influence he exerted on the young. It derived from the passion with which he had pursued his studies and later directed those of his pupils, while remaining indifferent to the honors and plaudits that are granted old Masters. (Levi-Montalcini 1988:205)

9. It must be stressed that drawing on insights of either Freud or Plato does not commit us to their doctrines. Diotima is still one of the great instructresses, not only of Plato's Socrates, but also of the Western tradition; that does not mean that in heeding her we are accepting the beautiful itself or the good itself or any of Plato's self-subsistent entities as real or explanatory. It is unfortunately the habit of many, especially biologists, to take any reference to Plato or Aristotle or Goethe or any non-Darwinian source as committing its authors to a vicious typology or "Platonism," whatever that is meant to mean. We wish to dissociate ourselves emphatically from any such supposed entailment.

References

Alexander, R. D. 1975. The search for a general theory of behavior. *Behav. Sci.* 20:77–100.

Allen, T. F. H. and T. B. Starr. 1982. *Hierarchy: Perspectives for Ecological Complexity.* Chicago: University of Chicago Press.

Andersson, M. 1984. The evolution of eusociality. *Ann. Rev. Ecol. Syst.* 15:165–89.

Antonovics, J. 1987. The evolutionary dys-synthesis: Which bottles for which wine? *Am. Nat.* 129:321–31.

Auffenberg, W. 1981. *The Behavioral Ecology of the Komodo Monitor.* Gainesville: University Presses of Florida.

Ayala, F. J. 1968. Biology as an autonomous science. *Amer. Sci.* 56:207–21.

Balme, D. M. 1962. Genos and Eidos in Aristotle's biology. *Class. Quart.* 56:91–104.

Barnes, R. D. 1963. *Invertebrate Zoology.* 1st ed. Philadelphia: W. B. Saunders.

Bell, G. 1982. *The Masterpiece of Nature.* Berkeley and Los Angeles: University of California Press.

Bendall, D. S., ed. 1983. *Evolution from Molecules to Men.* Cambridge: Cambridge University Press.

Bernstein, H., F. A. Hopf, and R. E. Michod. 1988. Is meiotic recombination an adaptation for repairing DNA, producing genetic variation, or both? In R. E. Michod and B. R. Levin, eds., *The Evolution of Sex*, pp. 139–60. Sunderland, Mass.: Sinauer Associates.

212   REFERENCES

Bloch, O. R. 1971. *La philosophie de Gassendi.* The Hague: Nijhoff.
Bock, W. J. 1979. The synthetic explanation of macroevolutionary change— a reductionistic approach. In J. H. Schwartz and H. B. Rollins, eds., *Models and Methodologies in Evolutionary Theory. Bull. Carnegie Mus. Nat. Hist.* 13:20–69.
Bock, W. J. 1986. Species concepts, speciation, and macroevolution. In K. Iwatsuki, P. H. Raven, and W. J. Bock, eds., *Modern Aspects of Species,* pp. 31–57. Tokyo: University of Tokyo Press.
Bock, W. J. and G. H. von Wahlert. 1965. Adaptation and the form-function complex. *Evolution* 19:269–99.
Bourdieu, P. 1977. *Outline of a Theory of Practice.* Cambridge: Cambridge University Press.
Bowler, P. J. 1983. *The Eclipse of Darwinism.* Baltimore: Johns Hopkins University Press.
Boyd, R. and P. J. Richerson. 1985. *Culture and the Evolutionary Process.* Chicago: University of Chicago Press.
Brandon, R. N. 1981. A structural description of evolutionary theory. *PSA 1980* 2:427–39.
Brandon, R. N. 1990. *Adaptation and Environment.* Princeton: Princeton University Press.
Bretsky, P. W. 1969. Evolution of Paleozoic benthic marine invertebrate communities. *Palaeogr. Palaeoclimatol. Palaeoecol.* 6:45–59.
Buchdahl, G. 1969. *Metaphysics and the Philosophy of Science.* Oxford: Blackwell.
Burian, R. M. and M. Grene. 1992. The philosophical debate in biology. *Enciclopedia Italiana: Storia del Secolo 20.*
Buss, Leo. 1987. *The Evolution of Individuality.* Princeton: Princeton University Press.
Cain, A. J. and P. M. Sheppard. 1950. Selection in the polymorphic land snail *Cepaea nemoralis* (L.). *Heredity* 4:275–94.
Caplan, A. L. and W. J. Bock. 1988. Haunt me no longer. *Biology and Philosophy* 3:443–54.
Carneiro, R. 1970. A theory of the origin of the state. *Science* 169:733–38.
Cassirer, E. 1933. Le language et la construction du monde des objets. *Jl. de Psychologie Normale et Pathologique* 30:18–44.
Cracraft, J. 1989. Species as entities of biological theory. In M. Ruse, ed., *What the Philosophy of Biology Is,* pp. 31–52. Dordrecht: Kluwer.
Cumberland, R. 1672. *De Legibus Naturae Disquisitio Philosophica.* London: N. Hooke.
Damuth, J. 1985. Selection among "species": A formulation in terms of natural functional units. *Evolution* 39:1132–46.
Darwin, C. 1859. *On the Origin of Species.* London: John Murray.
Darwin, C. 1862. *The Various Contrivances by Which Orchids Are Fertilized by Insects.* London: Murray.
Darwin, C. 1871. *The Descent of Man.* London: Murray.

Darwin, C. 1882. *Collected Life and Letters.* Edited by F. Darwin. 3 vols. London: Murray.

Darwin, C. 1958a. Sketch of 1842, Essay of 1844. In C. Darwin and A. R. Wallace, *Evolution by Natural Selection,* pp. 41–252. Cambridge: Cambridge University Press.

Darwin, C. 1958b. *Autobiography.* London: Collins.

Darwin, C. [1859] 1964. *Origin of Species.* Reprint. Cambridge: Harvard University Press.

Dawkins, R. 1976. *The Selfish Gene.* Oxford: Oxford University Press.

Dawkins, R. 1982. *The Extended Phenotype.* Oxford and San Francisco: Freeman.

Dawkins, R. 1986. *The Blind Watchmaker.* New York: Norton.

de Beer, G. 1958. The Darwin-Wallace centenary. *Endeavor* 17:61–76.

Depew, D. J. and B. H. Weber. 1988. Consequences of nonequilibrium thermodynamics for Darwinism. In B. H. Weber, D. J. Depew, and J. D. Smith, eds., *Entropy, Information and Evolution,* pp. 317–54. Cambridge: MIT Press.

Descartes, R. [1637] 1973a. *La Dioptrique.* In C. Adam and P. Tannery, eds., *Oeuvres,* vol. 6. Paris: Vrin.

Descartes, R. [1637] 1973b. *Discours de la Méthode.* In C. Adam and P. Tannery, eds., *Oeuvres,* vol. 6. Paris: Vrin.

Descartes, R. [1644] 1973c. *Principia Philosophiae.* In C. Adam and P. Tannery, eds., *Oeuvres,* vol. 8, 1. Paris: Vrin.

Descartes, R. [1632?]. *Le Monde.* In C. Adam and P. Tannery, eds., *Oeuvres,* vol. 11. Paris: Vrin.

Dobzhansky, T. 1937. *Genetics and the Origin of Species.* New York: Columbia University Press.

Dobzhansky, T. 1941. *Genetics and the Origin of Species.* 2d ed. New York: Columbia University Press.

Dobzhansky, T. 1968. On Cartesian and Darwinian aspects of biology. *Grad. Jl.* 8:99–117.

Ehrlich, P. R. and P. H. Raven. Differentiation of populations. *Science* 165:1228–32.

Eickwort, G. C. 1986. First steps into eusociality: The sweat bee *Dialictus lineatulus. The Florida Entomologist.* 69:742–54.

Eldredge, N. 1982. *The Monkey Business: A Scientist Looks at Creationism.* New York: Washington Square Books.

Eldredge, N. 1985. *Unfinished Synthesis.* New York: Oxford University Press.

Eldredge, N. 1986. Information, economics, and evolution. *Ann. Rev. Ecol. Syst.* 17:351–69.

Eldredge, N. 1989. *Macroevolutionary Dynamics.* New York: McGraw-Hill.

Eldredge, N. 1992. Species, selection, and Paterson's concept of the Specific Mate Recognition System. In D. Lambert, H. Spencer, and J. Masters, eds., *Speciation and the Recognition Concept: Development, Theory, and Application.* Baltimore: Johns Hopkins University Press.

Eldredge, N. In press. The biology of social systems. In E. Khalil and K. Boulding, eds., *Social and Natural Complexity.* New York: Cambridge University Press.

Eldredge, N. and S. J. Gould. 1972. Punctuated equilibria: An alternative to phyletic gradualism. In T. J. M. Schopf, ed., *Models in Paleobiology,* pp. 82–115. San Francisco: Freeman, Cooper.

Eldredge, N. and S. N. Salthe. 1984. Hierarchy and evolution. *Oxford Survs. Evol. Biol.* 1:182–206.

Emlen, J. 1988. Evolutionary ecology and the optimality assumption. In J. Dupré, ed., *The Latest on the Best: Essays on Evolution and Optimality.* Cambridge: MIT Press.

Endler, J. 1986. *Natural Selection in the Wild.* Princeton: Princeton University Press.

Endler, J. A. and T. McClellan. 1988. The processes of evolution: Toward a newer synthesis. *Ann. Rev. Ecol Syst.* 19:395–422.

Fisher, H. 1982. *The Sex Contract.* New York: W. Morrow.

Fisher, R. A. [1930] 1958. Reprint. *The Genetical Theory of Natural Selection.* New York: Dover.

Futuyma, D. 1983. *Science on Trial.* New York: Pantheon.

Galilei, G. 1957. *Discoveries and Opinions of Galileo.* Translated by S. Drake. New York: Doubleday.

Gallup, G. G. 1970. Chimpanzees: Self-recognition. *Science* 167:86–87.

Geertz, C. 1988. *Works and Lives: The Anthropologist as Author.* Stanford: Stanford University Press.

Ghiselin, M. T. 1974a. A radical solution to the species problem. *Syst. Zool.* 23:536–44.

Ghiselin, M. T. 1974b. *The Economy of Nature and the Evolution of Sex.* Berkeley and Los Angeles: University of California Press.

Ghiselin, M. T. 1987. Species concepts, individuality, and objectivity. *Biol. and Phil.* 2:127–43.

Goldschmidt, R. 1940. *The Material Basis of Evolution.* New Haven: Yale University Press.

Goodall, J. 1986. *The Chimpanzees of Gombe.* Cambridge: Harvard University Press.

Goodwin, B. C. 1988. Morphogenesis and heredity. In M.-W. Ho and S. W. Fox, eds., *Evolutionary Processes and Metaphors,* pp. 145–62. New York: Wiley.

Gordon, D. M. 1988. Behaviour changes—finding the rules. In M.-W. Ho and S. W. Fox, eds., *Evolutionary Processes and Metaphors,* pp. 243–54. New York: Wiley.

Gordon, D. M. 1990. Caste and change in social insects. *Oxford Surveys in Evolutionary Biology* 6:55–72.

Gould, S. J. 1977. *Ontogeny and Phylogeny.* Cambridge: Harvard University Press.

Gould, S. J. 1980a. Is a new and general theory of evolution emerging? *Paleobiol.* 6:119–30.

Gould, S. J. 1980b. G. G. Simpson, paleontology, and the modern synthesis. In E. Mayr and W. B. Provine, eds., *The Evolutionary Synthesis: Perspectives on the Unification of Biology,* pp. 153–72. Cambridge: Harvard University Press.

Gould, S. J. 1983. The hardening of the modern synthesis. In M. Grene, ed., *Dimensions of Darwinism,* pp. 71–93. Cambridge: Cambridge University Press.

Gould, S. J. 1986. Evolution and the triumph of homology, or why history matters. *Amer. Sci.* 74:60–69.

Gould, S. J. and R. C. Lewontin. 1979. The spandrels of San Marco and the Panglossian paradigm: A critique of the adaptationist programme. *Proc. Roy. Soc. London, B.* 205:581–98.

Gould, S. J. and E. S. Vrba. 1982. Exaptation—a missing term in the science of form. *Paleobiology* 8:4–15.

Grant, P. R. 1986. *Ecology and Evolution of Darwin's Finches.* Princeton: Princeton University Press.

Gregory, M. S., A. Silvers, and D. Sutch, eds. 1978. *Sociobiology and Human Nature.* San Francisco: Jossey-Bass.

Grene, M. 1959. Two evolutionary theories. *British Journal for the Philosophy of Science* 9:110–27, 185–93.

Grene, M. 1964. *A Portrait of Aristotle.* Chicago: University of Chicago Press.

Grene, M. 1974. *The Understanding of Nature.* Dordrecht: Reidel.

Grene, M. 1978. Sociobiology and the human mind. In M. S. Gregory, A. Silvers, and D. Sutch, eds., *Sociobiology and Human Nature,* pp. 213–24. San Francisco: Jossey-Bass.

Grene, M. 1981. Changing concepts of Darwinian evolution. *The Monist.* 64:195–213.

Grene, M., ed. 1983. *Dimensions of Darwinism.* Cambridge: Cambridge University Press.

Grene, M. 1987. Hierarchies in biology. *American Scientist* 75:504–10.

Grene, M. 1989. Interaction and evolution. In M. Ruse, ed., *What the Philosophy of Biology Is,* pp. 67–73. Dordrecht: Kluwer Academic Publishers.

Grene, M. 1989. Animal mechanism and the Cartesian vision of nature. Manuscript.

Hagen, J. B. 1989. Research perspectives and the anomalous status of modern ecology. *Biology and Philosophy* 4:433–55.

Hamilton, W. D. 1964a. The genetical evolution of social behavior, 1. *Jl. Theor. Biol.* 7:1–16.

Hamilton, W. D. 1964b. The genetical evolution of social behavior, 2 *Jl. Theor. Biol.* 7:17–52.

Hamilton, W. D. 1975. Innate social aptitudes of man: An approach from

216   REFERENCES

evolutionary genetics. In R. Fox, ed., *Biosocial Anthropology*, pp. 37–67.
London: Malaby Press.

Harper, J. L. 1983. A Darwinian plant ecology. In D. S. Bendall, ed.,
Evolution from Molecules to Men, pp. 323–45. Cambridge: Cambridge
University Press.

Harris, M. 1989. *Our Kind*. New York: Harper and Row.

Hawkins, H. L. 1953. Fossils and men. *Sci. Jl. Royal Coll. Sci.* 23:1–10.

Hillis, D. M. 1988. Systematics of the *Rana pipiens* complex: Puzzle and
paradigm. *Ann. Rev. Ecol. Syst.* 19:39–64.

Hillis, D. M., J. S. Frost, and D. A. Wright. 1983. Phylogeny and biogeography of the *Rana pipiens* complex: A biochemical evaluation. *Syst. Zool.*
32:132–43.

Hodge, M. J. S. 1977. The structure and strategy of Darwin's "long argument." *Brit. Jl. Hist. Sci.* 10:237–46.

Hull, D. L. 1965. The effects of essentialism on taxonomy: Two thousand
years of stasis. *Brit. Jl. Phil. Sci.* 15:314–26; 16:1–18.

Hull, D. L. 1976. Are species really individuals? *Syst. Zool.* 25:174–91.

Hull, D. L. 1978. A matter of individuality. *Phil. Sci.* 45:335–60.

Hull, D. L. 1980. Individuality and selection. *Ann. Rev. Ecol. Syst.*
11:311–32.

Hume, D. [1739] 1975. *A Treatise of Human Nature*. Reprint. Oxford: Oxford University Press.

Imbrie, J. 1959. Brachiopods of the Traverse Group (Devonian) of Michigan. Part 1: Dalmanellacea, Pentameracea, Strophomenacea, Ortho-
tetacea, Chonetacea, and Productacea. *Bull. Amer. Mus Nat. Hist.*
116(4):345–410.

Jacob, F. 1970. *The Possible and the Actual*. Seattle: University of Washington
Press.

Kant, I. [1755] 1968. Allgemeine Naturgeschichte und Theorie des Him-
mels. In *Kants Werke, Akademie Ausgabe*, vol. 1, pp. 215–368. Berlin: De
Gruyter.

Kant, I. [1787] 1968. Kritik der reinen Vernunft. 2d ed. In *Kants Werke,
Akademie Ausgabe*, vol. 3. Berlin: De Gruyter.

Kauffman, S. 1985. Self-organization, selective adaptation and its limits: a
new pattern of inference in evolution and development. In D. J. Depew
and B. H. Weber, eds., *Evolution at a Crossroads*, pp. 169–208. Cambridge: MIT Press.

Kauffman, S. 1992. *Origins of Order: Self-Organization and Selection in Evolution*. Oxford: Oxford University Press.

Kevles, D. J. 1985. *In the Name of Eugenics*. New York: Knopf.

Kirkpatrick, M. 1982. Sexual selection and the evolution of female choice.
Evolution 36:1–12.

Kitcher, P. 1982. *Abusing Science*. Cambridge: MIT Press.

Kitcher, P. 1984. 1953 and all that: A tale of two sciences. *Phil. Rev.*
93:335–73.

Kitcher, P. 1985a. Darwin's achievement. In N. Rescher, ed., *Reason and Rationality in Science*, pp. 127–89. Washington, D.C.: University Press of America.

Kitcher, P. 1985b. *Vaulting Ambition*. Cambridge: MIT Press.

Krebs, J. R. and N. B. Davies. 1984. *Behavioural Ecology: An Evolutionary Approach*. 2d ed. Sunderland, Mass.: Sinauer Associates.

Krebs, J. R. and R. H. McLeery. 1984. Optimization in behavioural ecology. In J. R. Krebs and N. B. Davies, eds., *Behavioural Ecology: An Evolutionary Approach*, pp. 91–121. Sunderland, Mass.: Sinauer Associates.

Lamotte, M. 1959. Polymorphism of natural populations of *Cepaea nemoralis. Cold Spring Harbor Symposia on Quant. Biol.* 24:65–86.

Lande, R. 1980. Sexual dimorphism, sexual selection, and adaptation in polygenic characters. *Evolution* 34:292–305.

Lande, R. 1981. Models of speciation by sexual selection on polygenic characters. *Proc. Nat. Acad. Sci.* 78:3721–25.

Lee, J. J. and P. Hallock. 1987. Algal symbiosis as the driving force in the evolution of larger foraminifera. In J. J. Lee and J. Frederick, eds., *Endosymbiosis 3. Ann. N.Y. Acad. Sci.* 503:330–47.

Lenoir, T. 1989. *The Strategy of Life*. Chicago: University of Chicago Press.

Levi-Montalcini, R. 1988. *In Praise of Imperfection*. New York: Basic Books.

Levins, R. 1966. The strategy of model building in population biology. *Amer. Sci.* 54:421–31.

Levinton, J. 1988. *Genetics, Paleontology, and Macroevolution*. Cambridge and New York: Cambridge University Press.

Lewontin, R. C. 1983. Gene, organism, and environment. In D. S. Bendall, *Evolution from Molecules to Men*, pp. 273–86. Cambridge: Cambridge University Press.

Locke, J. [1690] 1975. *An Essay Concerning Human Understanding*. Reprint, edited by P. Nidditch. Oxford: Oxford University Press.

Löwith, K. 1967. *Gott, Mensch und Welt in der Metaphysik von Descartes bis zu Nietzsche*. Göttingen: Vandenhoeck and Ruprecht.

Malthus, T. R. [1798] 6th ed. 1826. *An Essay on the Principle of Population*. London: Murray.

Margulis, L. 1970. *Origin of Eukaryotic Cells*. New Haven: Yale University Press.

Margulis, L. 1974. Five-kingdom classification and the origin and evolution of cells. *Evol. Biol.* 7:45–78.

Margulis, L. 1981. *Symbiosis in Cell Evolution: Life and Environment on the Early Earth*. San Francisco: Freeman.

Margulis, L. and D. Sagan. 1986. *Microcosmos*. New York: Summit.

Maynard Smith, J. 1977. Parental investment—a prospective analysis. *Animal Behavior* 25:1–9.

Maynard Smith, J. 1978. *The Evolution of Sex*. Cambridge: Cambridge University Press.

Maynard Smith, J., R. M. Burian, S. Kauffman, P. Alberch, J. Campbell,

B. Goodwin, R. Lande, D. Raup, and L. Wolpert. 1985. Developmental constraints and evolution. *Quart. Rev. Biol.* 60:265–87.

Mayr, E. 1942. *Systematics and the Origin of Species.* New York: Columbia University Press.

Mayr, E. 1970. *Populations, Species and Evolution.* Cambridge: Harvard University Press.

Mayr, E. 1974. Teleological and teleonomic: A new analysis. *Boston Studies in Philosophy of Science.* 14:91–117.

Mayr, E. 1978. Evolution. In *Evolution*, pp. 3–9. Freeman: San Francisco.

Mayr, E. 1982. *The Growth of Biological Thought.* Cambridge: Harvard University Press.

Mayr, E. 1983. How to carry out the adaptationist program. *Amer. Nat.* 121:324–33.

Mayr, E. 1988. *Toward a New Philosophy of Biology.* Cambridge: Harvard University Press.

Mayr, E. and W. B. Provine, eds. 1980. *The Evolutionary Synthesis.* Cambridge: Harvard University Press.

Merlau-Ponty, M. 1942. *La structure du comportement.* Paris: Universitaires de France.

Michod, R. E., and B. R. Levin, eds. 1988. *The Evolution of Sex.* Sunderland, Mass.: Sinauer Associates.

Mishler, B. 1988. Criticism of two hierarchy views. Ms.

Mishler, B. D. and R. N. Brandon. 1987. Individuality, pluralism, and the phylogenetic species concept. *Biol. & Phil.* 2:397–414.

Monod, J. 1970. *Le hasard et la nécessité.* Paris: Editions du Seuil.

Moore, J. A. 1949. Patterns of evolution of the genus *Rana.* In G. L. Jepsen, G. G. Simpson, and E. Mayr, eds., *Genetics, Paleontology and Evolution,* pp. 315–38. Princeton: Princeton University Press.

Moore, J. A. 1975. *Rana pipiens*—the changing paradigm. *Am. Zool.* 15:837–49.

Morton, J. E. 1958. *Molluscs.* London: Hutchinson and Co.

Moss, C. 1988. *Elephant Memories: Thirteen Years in the Life of an Elephant Family.* New York: W. Morrow.

Munn, N. 1986. *The Fame of Gawa: A Symbolic Study of Value Transformation in a Massim (Papua, New Guinea) Society.* Cambridge: Cambridge University Press.

Newell, N. D. 1982. *Creation and Evolution: Myth or Reality?* New York: Columbia University Press.

Newton, I. 1962. *Mathematical Principles.* Cajori revision of Motte translation, vol. 2. Berkeley and Los Angeles: University of California Press.

O'Grady, R. T. 1984. Evolutionary theory and teleology. *J. Theor. Biol.* 107:563–78.

Ortner, S. 1984. Theory in anthropology since the sixties. *Comp. Stud. Soc. Hist.* 26:126–65.

Oster, G. F. and E. O. Wilson. 1978. *Caste and Ecology in the Social Insects.* Princeton: Princeton University Press.

Paley, W. 1802. *Natural Theology: Or, Evidence of the Existence and Attributes of the Deity Collected from the Appearances of Nature.* London: R. Fauldner.

Passingham, R. 1982. *The Human Primate.* San Francisco: Freeman.

Paterson, H. E. H. 1985. The recognition concept of species. In E. S. Vrba, ed., *Species and Speciation, Transvaal Mus. Monogr.* 4:21–29.

Pellegrin, P. 1986. *Aristotle's Classification of Animals.* Berkeley and Los Angeles: University of California Press.

Pico della Mirandola. 1956. *Treatise on Man.* Chicago: Regnery.

Pilette, R. 1989. *Community Relationships of the Hamilton Fauna of New York State: An Application of Community and Hierarchical Models and Theories to Paleoecology.* Ph.D. diss., City University of New York.

Plessner, H. 1928. *Die Stufen des Organischen und der Mensch.* Berlin: De Gruyter.

Portmann, A. 1956. *Zoologie und des neue Bild des Menschen.* Hamburg: Rowohlt.

Provine, W. B. 1971. *The Origins of Theoretical Population Genetics.* Chicago: University of Chicago Press.

Provine, W. B. 1986. *Sewall Wright and Evolutionary Biology.* Chicago: University of Chicago Press.

Provine, W. B. 1986. Progress in evolution and meaning in life. In M. Nitecki, ed., *Evolutionary Progress,* pp. 49–74. Chicago: University of Chicago Press.

Reed, E. 1988. *James J. Gibson and the Psychology of Perception.* New Haven: Yale University Press.

Richard, A. 1985. *Primates in Nature.* San Francisco: Freeman.

Ridley, M. 1989. The cladistic solution to the species problem. *Biol. & Phil.* 4:1–16.

Robson, G. C. and O. W. Richards. 1936. *The Variation of Animals in Nature.* London: Longmans Green.

Roger, J. 1963. *Les sciences de la vie dans la pensée française de la XVIIIe siècle.* Paris: Armand Colin.

Rohault, J. 1671a. *Entretiens sur la philosophie.* Paris: M. le Petit.

Rohault, J. 1671b. *Traité de physique.* Paris: Vve. de C. Savreux.

Rousseau, J. J. [1762] 1973. *The Social Contract.* Reprint. London: Dent.

Rubenstein, D. I. and R. W. Wrangham. 1986. Socioecology: Origins and trends. In D. I. Rubenstein and R. W. Wrangham, eds., *Ecological Aspects of Social Evolution: Birds and Mammals,* pp. 3–17. Princeton: Princeton University Press.

Rubenstein, D. I. and R. W. Wrangham, eds. 1986. *Ecological Aspects of Social Evolution: Birds and Mammals.* Princeton: Princeton University Press.

Rumbaugh, D. 1985. Comparative psychology: Patterns in adaptation. In A. M. Rogers and C. J. Scheirer, eds., *The G. Stanley Hall Lecture Series* 5, pp. 7–53. Washington, D.C.: Amer. Psych. Assoc.

Ruse, M. 1986. *Taking Darwin Seriously: A Naturalistic Approach to Philosophy.* Oxford: Blackwell.

Ryan, M. J. 1985. *The Tungara Frog.* Chicago: University of Chicago Press.

Ryan, M. J. and W. Wilczynski. 1988. Coevolution of sender and receiver: Effect on local mate preference in cricket frogs. *Science* 240:1786–88.

Sahlins, M. 1976. *The Use and Abuse of Biology.* Ann Arbor: University of Michigan Press.

Salthe, S. N. 1985. *Evolving Hierarchical Systems.* New York: Columbia University Press.

Savage-Rumbaugh, E. S. and B. J. Wilkerson. 1978. Socio-sexual behavior in *Pan paniscus* and *Pan tryglodytes:* A comparative study. *Jl. Human Evol.* 7:327–44.

Savage-Rumbaugh, S., D. Rumbaugh, and K. McDonald. 1985. Language learning in two species of apes. *Neurosci. and Biobehav. Revs.* 9:653–65.

Schaeffer, B. 1965. The role of experimentation in the origin of higher levels of organization. *Systematic Zool.* 14:318–36.

Searle, J. R. 1978. Sociobiology in the evolution of behavior. In M. S. Gregory, A. Silvers, and D. Sutch, eds., *Sociobiology and Human Nature,* pp. 164–82. San Francisco: Jossey-Bass.

Shapin, S. and S. Schaffer. 1985. *Leviathan and the Air Pump.* Princeton: Princeton University Press.

Shapiro, R. 1986. *Origins.* New York: Summit Books.

Sherman, P. W., J. U. M. Jarvis, and R. D. Alexander, eds. 1991. *The Biology of the Naked Mole-rat.* Princeton: Princeton University Press.

Simberloff, D. S. 1972. Models in biogeography. In T. J. M. Schopf, ed., *Models in Paleobiology,* pp. 160–91. San Francisco: Freeman, Cooper.

Simon, H. A. 1962. The architecture of complexity. *Proc. Am. Phil. Soc.* 106:467–82.

Simpson, G. G. 1944. *Tempo and Mode in Evolution.* New York: Columbia University Press. Reprinted, 1984.

Simpson, G. G. 1953. *The Major Features of Evolution.* New York: Columbia University Press.

Simpson, G. G. 1961. *Principles of Animal Taxonomy.* New York: Columbia University Press.

Smith, J. D. 1985. *Minds Made Feeble: The Myth and Legacy of the Kallikaks.* Rockville, Md.: Aspen.

Sokal, R. R. and T. J. Crovello. 1970. The biological species concept: A critical evaluation. *Amer. Naturalist.* 104:127–53.

Stanley, S. M. 1985. Rates of evolution. *Paleobiology* 11:13–26.

Stebbins, G. L. 1987. Lecture at Cornell University.

Stebbins, G. L. and F. J. Ayala. 1981. Is a new evolutionary synthesis necessary? *Science* 213:967–71.

Steele, E. J. 1979. *Somatic Selection and Adaptive Evolution.* Toronto: Williams and Wallace.

Stenseth, N. C. and J. Maynard Smith. 1984. Coevolution in ecosystems: Red Queen evolution or stasis? *Evolution* 38:870–80.

Stephens, D. W. and J. R. Krebs. 1986. *Foraging Theory*. Cambridge: Cambridge University Press.

Tattersall, I. 1982. *The Primates of Madagascar*. New York: Columbia University Press.

Tax, Sol. 1960. *Evolution After Darwin*. Vols. 1–3. Chicago: University of Chicago Press.

Taylor, C. 1989. *Sources of the Self*. Cambridge: Harvard University Press.

Tinbergen, N. 1960. *The Herring Gull's World*. New York: Basic Books.

Ulanowicz, R. E. 1986. *Growth and Development: Ecosystems Phenomenology*. New York: Springer Verlag.

Van Valen, L. 1973. A new evolutionary law. *Evol. Theory* 1:1–30.

Vrba, E. S. 1985. Introductory comments on species and speciation. In E. S. Vrba, ed., *Species and Speciation, Transvaal Mus. Monogr.* 4:ix–xviii.

Vrba, E. S. 1988. Ecological predictions for macroevolutionary patterns in the fossil record. In N. C. Stenseth, ed., *Coevolution in Ecosystems and the Red Queen Hypothesis*. Cambridge: Cambridge University Press.

Vrba, E. S. 1989. Levels of selection and sorting with special reference to the species level. *Oxford Survs. in Evol. Biol.* 6:111–68.

Vrba, E. S. and N. Eldredge. 1984. Individuals, hierarchies, and processes: Towards a more complete evolutionary theory. *Paleobiology* 10:146–71.

Vrba, E. S. and S. J. Gould. 1986. The hierarchical expansion of sorting and selection: Sorting and selection cannot be equated. *Paleobiology* 12:217–28.

Washburn, S. L. 1978. Animal behavior and social anthropology. *In* M. S. Gregory, A. Silvers, and D. Sutch, eds., *Sociobiology and Human Nature*, pp. 53–74. San Francisco: Jossey-Bass.

Weber, B. H., D. J. Depew, and J. D. Smith. 1988. *Entropy, Information and Evolution*. Cambridge: MIT Press.

Webster, G. 1984. The relations of natural forms. In M.-W. Ho and P. Saunders, eds., *Beyond Neo-Darwinism*, pp. 193–217. New York: Academic Press.

Webster, G. and Goodwin, B. C. 1982. The origin of species: A structuralist approach. *Jl. Soc. Biol. Struc.* 5:15–47.

West-Eberhard, M. J. 1979. Sexual selection, social competition and evolution. *Proc. Amer. Phil. Soc.* 123:222–34.

Westfall, W. S. 1977. *The Construction of Modern Science: Mechanism and Mechanics*. Cambridge: Cambridge University Press.

Westfall, W. S. 1980. *Never at Rest*. Cambridge: Cambridge University Press.

Whittaker, R. H. 1969. New concepts of kingdoms of organisms. *Science* 163:150–60.

Wiley, E. O. 1978. The evolutionary species concept reconsidered. *Syst. Zool.* 27:17–26.

Wiley, E. O. 1981. *Phylogenetics: The Theory and Practice of Phylogenetic Systematics*. New York: Wiley.

Williams, G. C. 1966. *Adaptation and Natural Selection.* Princeton: Princeton University Press.

Williams, G. C. 1975. *Sex and Evolution.* Princeton: Princeton University Press.

Willis, J. C. 1940. *The Course of Evolution.* Cambridge: Cambridge University Press.

Wilson, E. O. 1975. *Sociobiology.* Cambridge: Harvard University Press.

Wilson, E. O. 1978. *On Human Nature.* Cambridge: Harvard University Press.

Wilson, E. O. 1985. The sociogenesis of insect colonies. *Science* 228: 1489–95.

Wilson, P. 1980. *Man: The Promising Primate.* New Haven: Yale University Press.

Wimsatt, W. C. 1980. Reductionistic research strategies and their biases in the units of selection controversy. In T. Nickles, ed., *Scientific Discovery: Case Studies,* pp. 213–59. Dordrecht: D. Reidel.

Winkler, D. W. and G. S. Wilkinson. 1988. Parental effort in birds and mammals: Theory and measurement. *Oxford Surveys in Evolutionary Biology* 5:185–214.

Winston, J. E. 1986. Victims of avicularia. *P. S. Z. N. I: Marine Ecology* 7:193–99.

Winston, J. E. 1990. Intertidal space wars. *Sea Frontiers* 36:46–51.

Woese, C. R. 1981. Archaebacteria. *Scientific American* 244:98–122.

Woolfenden, G. E. and J. W. Fitzpatrick. 1984. *The Florida Scrub Jay: Demography of a Cooperative-Breeding Bird.* Princeton: Princeton University Press.

Wrangham, R. W. 1986. Ecology and social relationships in two species of chimpanzees. In D. I. Rubenstein and R. W. Wrangham, eds., *Ecological Aspects of Social Evolution: Birds and Mammals,* pp. 352–78. Princeton: Princeton University Press.

Wright, S. 1931. Evolution in Mendelian populations. *Genetics* 16:97–159.

Wright, S. 1932. The roles of mutation, inbreeding, crossbreeding, and selection in evolution. *Proc. Sixth Int. Congr. Genetics* 1:356–66.

Wright, S. 1967. Comments on the preliminary working papers of Eden and Waddington. In *Mathematical Challenges to the Neo-Darwinian Interpretation of Evolution,* pp. 117–20. Philadelphia: Wistar Institute Press.

Wynne-Edwards, V. C. 1962. *Animal Dispersion in Relation to Social Behaviour.* New York: Hafner.

Wynne-Edwards, V. C. 1965. Self-regulating systems in populations of animals. *Science* 147:1543–48.

Wynne-Edwards, V. C. 1986. *Evolution Through Group Selection.* Boston: Blackwell Scientific Publications.

Zahavi, A. 1974. Communal nesting by the Arabian Babbler: A case of individual selection. *Ibis* 116:84–87.

Index

Perspectivism, 175
Phenome (term), 207n7
Phenotype(s), 49, 59, 85, 102; features, functions, roles of, 60–61, 65
Phenotypic adaptations, 76–78
Phenotypic attributes/traits, 59, 60–61, 65, 69, 89, 96; economic success resulting from variations in, 63; in reproductive success, 61–62; species-specific, 205n7
Phenotypic change, 86–87
Phenotypic diversity, 87
Phenotypic selection, 12
Phenotypic stability, 86–87
Phenotypic variation, 50
Philosophical anthropology, 177, 183, 184–88, 189
Philosophy of science, 10, 50, 168
Phylogenesis, 124
Phylogenetic history, 101, 114, 127; and division of labor, 133–34
Phylogenetic markers, 104, 105
Phylogenetic novelty, 95
Phylogenetic relationships, 132
Phylogenetic species concepts, 26, 27
"Phylogenetic stream," 95
Phylogenetic systems, 125
Phylogeny, 168; of sociality, 147
Physalia, 137
Physical proportions: in human development, 181–82
Physics, 22–23, 33, 50, 168; atomistic, 34–35; Cartesian, 32, 35, 42; mechanistic, 40; new, 42; Newtonian, 32
Pico della Mirandola, 172–73, 189
Pilette, R., 114, 206–7n7
Plantae/plants, 37, 72, 108, 207n1; economic differentiation, 72; higher, 51–52; positionality of, 184
Plastids, 51
Plato, 199
Plessner, H., 183, 184–86, 189
Polyps, 136, 141, 144, 147, 151, 162
Population control, 192

Population genetics, 22, 56, 80, 90, 126; fitness in, 61
Populations, 51, 97; and economics, 112–13
Population size, 113, 163, 206n2; energy resources limiting, 110
Population thinking, 48
Portmann, A., 180, 181, 182, 183
Positionality (concept), 184, 186
Potential interbreeding, 91, 93
Precocial species, 180, 181
Predator-prey cycles, 113
Predators/predation, 11, 37, 206n2
Prey, 11, 37, 109
Primate communication, 170
Primates, 168, 173, 179; social behavior, 155–59, 170
Principia (Newton), 45
Principles (Descartes), 34, 35
Processes, 76–77; in organic realm, 59–78
Progress: in evolution, 46–47
Prokaryota/prokaryotes, 69, 70, 71–72, 133, 134; defining properties of, 70t
Prokaryotic cells, 66, 67
Properties (of species), 82–83
Protection, 157–58, 160–61, 208n8
Proteins, 66–68, 71
Protista, 72
Protostomes, 137
Provine, W. B., 21, 22, 42, 90
Pseudoteleology, 17–18, 19, 55
Pygmy chimps (*Pan paniscus*), 172, 174, 210n5

Qualities, primary/secondary, 36–37
Quantification, 10, 42

Racial senescence, 21
Random variation, 12, 54–55
Raven, P. H., 93, 94
Ray, J., 37
Recognition, 103, 250n7; in economic contexts, 112; *see also* Mate recognition
"Red Queen," 118